IntechOpen

Supporting open minds since 2005

Control Systems in Engineering and Optimization Techniques
http://dx.doi.org/10.5772/intechopen.95665
Edited by P. Balasubramaniam, Sathiyaraj Thambiayya, Kuru Ratnavelu and JinRong Wang

Contributors
Norizarina Ishak, Muhammad Jaffar Sadiq Abdullah, Pritee Nivrutti Hulule, Yossi Peretz, Vladimir L. Kodkin, Alexandr S. Anikin, Alexandr A. Baldenkov, Natalia A. Loginova, Kannan Nilakantan, Sathiyaraj Thambiayya, P. Balasubramaniam, JinRong Wang, Kuru Ratnavelu

First published in London, United Kingdom, 2022 by IntechOpen
IntechOpen is the global imprint of INTECHOPEN LIMITED, registered in England and Wales, registration number: 11086078, 5 Princes Gate Court, London, SW7 2QJ, United Kingdom
Printed in Croatia

British Library Cataloguing-in-Publication Data
A catalogue record for this book is available from the British Library

Additional hard and PDF copies can be obtained from orders@intechopen.com

Control Systems in Engineering and Optimization Techniques
Edited by P. Balasubramaniam, Sathiyaraj Thambiayya, Kuru Ratnavelu and JinRong Wang
p. cm.
Print ISBN 978-1-83969-788-3
Online ISBN 978-1-83969-789-0
eBook (PDF) ISBN 978-1-83969-790-6

For EU product safety concerns:
IN TECH d.o.o., Prolaz Marije Krucifikse Kozulić 3, 51000 Rijeka, Croatia, info@intechopen.com or visit our website at intechopen.com.

Meet the editors

P. Balasubramaniam received Ph.D. in Mathematics in 1994. He published nearly 260 research articles with H-index of 50. From the Gandhigram Rural University, India, he received many awards and recognitions.

Sathiyaraj Thambiayya was awarded Ph.D. in 2017 and subsequently completed post-doctorate in 2021 at Guizhou University, China. He published more than 22 research papers, serving as an editorial board member and reviewer of a few journals.

Kuru Ratnavelu served as a Deputy Vice-Chancellor (Development), at the University of Malaya, other positions including Deputy-Dean and Head. He received Ph.D. in atomic physics (1990). He published nearly 200 articles.

JinRong Wang received a Ph.D. degree in 2009 at Guizhou University, China. He published more than 220 articles with an H-index of 51 in the area of Control Theory.

Contents

Preface

Applications of many mathematical modeling lead us to study the qualitative behaviors, including the controllability/stability of the systems. It is one of the most sought interdisciplinary areas of research and arises in the very first technological discoveries of the Industrial Revolution and modern technological applications. Control theory is an important discipline. It emerged in portfolio diversification strategy for the selected stocks and applications in stability analysis, supply chain dynamics, TCP strategic planning, electrical plant distribution systems, and the stabilization of airplanes.

In this collection of chapters in *Control Systems in Engineering and Optimization Techniques*, we have selected various multi-disciplinary research fields that fall within the ambit of Control Theory and applications. The reader will be given insights on these varieties of topics ranging from financial markets, supply chain dynamics, test case prioritization, static output feedback and applications, experimental study of electric drives, and stability issues in fractional differential systems.

In chapter 1, the best portfolio diversification strategy for the selected stocks from the Tokyo Stock Exchange had been examined using the Markowitz mean value method during different subperiods of time, which revealed several interesting insights. The naïve diversification strategy is used to serve as a comparison for the approach used. Performance-wise the optimal portfolio dominated the naïve strategy throughout the subperiods tested. The stability of fractional differential equations (FDEs) has shown wide applications in various sciences and engineering fields. Chapter 2 deals with the problem of finite-time stability of fractional higher-order stochastic delay systems. The chapter provides compelling ideas about the strength of stochastic fractional stability. Using a solution representation given by using sine-cosine functions for various delay intervals, various existence and uniqueness results are proved through the fixed point theorem. Further, finite-time stability criteria were demonstrated using a fractional Gronwall-Bellman inequality lemma.

Supply chain dynamics essentially deal with the dynamic behavior and temporal variation of the inventories and flows in the system over time when subjected to demand disturbances. In chapter 3, the discrete-time control supply chains problem is studied. Direct operator methods were used instead of the conventional block diagrams and the transfer function theory. The disturbance induced by demand deviation from the planned/anticipated levels was considered by considering the inventory level. Testing software or application is a fundamental part of SDLC. In chapter 4 a novel approach to Test Case Prioritization (TCP) and Strategies are developed and presented.

In chapter 5, an explicit free parametrization is presented for all the stabilizing static-state-feedbacks for continuous-time LTI systems. The parametrizations are utilized to derive optimal control results, pole placement, and exact pole assignment problems. Chapter 6 deals with the results of experiments in which the modes of parrying step loads are represented by the maximum possible recorded signals-frequency and amplitude of the stator voltage generated by various algorithms, the frequency and amplitude of the rotor current, the natural slip, and the rotational speed of the motor rotor. This made it possible to assess the effectiveness of interpreting asynchronous electric drives and methods of their regulation as objectively as possible. Over the past 25-30 years, numerous articles on this topic have not provided such results.

The book can be used as a reference book for several interdisciplinary research areas. We express our sincere thanks to all the contributors, managing editors, and typesetters.

P. Balasubramaniam
Professor,
Department of Mathematics,
Gandhigram Rural Institute,
Gandhigram, Tamil Nadu, India

Dr. Sathiyaraj Thambiayya
Institute of Actuarial Science and Data Analytics,
UCSI University,
Kuala Lumpur, Malaysia

Kuru Ratnavelu
Professor,
Institute of Computer Science and Digital Innovation,
UCSI University,
Kula Lumpur, Malaysia

JinRong Wang
Department of Mathematics,
Guizhou University,
Guiyang, China

Section 1

Controls and Fractional Order Systems

Chapter 1

An Optimal Control Approach to Portfolio Diversification on Large Cap Stocks Traded in Tokyo Stock Exchange

Muhammad Jaffar Sadiq Abdullah and Norizarina Ishak

Abstract

In this chapter, Markowitz mean-variance approach is proposed for examining the best portfolio diversification strategy within three subperiods which are during the global financial crisis (GFC), post-global financial crisis, and during the non-crisis period. In our approach, we used 10 securities from five different industries to represent a risk-mitigation parameter. In this way, the naive diversification strategy is used to serve as a comparison for the approach used. During the computation process, the correlation matrices revealed that the portfolio risk is not well diversified during non-crisis periods, meanwhile, the variance-covariance matrices indicated that volatility can be minimized during portfolio construction. On this basis, 10 efficient portfolios were constructed and the optimal portfolios were selected in each subperiods based on the risk-averse preference. Performance-wise that optimal portfolio dominated the naïve strategy throughout the three subperiods tested. All the optimal portfolios selected are yielding more returns compared to the naïve portfolio.

Keywords: naïve diversification, mean-variance, Sharpe ratio, efficient frontier, portfolio optimization

1. Introduction

Investment decision and capital allocation in the stock market is subjected to the variability of risk and return. Thus, the investors, as well as the portfolio manager, must find the best solutions to allocate the various assets efficiently. Constructing a portfolio of a stock based on the investor's risk tolerance is a difficult task. In this situation, the portfolio manager needs to have some background knowledge in the economy of the market and mathematical modeling to create a portfolio based on the market participant's appetite.

In the year 1952, Markowitz's has introduced mean-variance portfolio optimization, where the investors are called rational investors by analyzing the mean and variance to determine the value of expected return and risk preference for investors [1]. This modern portfolio theory suggested that an investor can perform diversification in allocating assets in their portfolios by concerning how much risk they are willing to bear due to the outcome uncertainty in the stock market.

On the risk-return spectrum, some of the investor's favor seeking an opportunity to invest in the large-cap stock due to its stability and safer investment during turbulent times. In this study, the Tokyo Stock Exchange (TSE) is chosen as a medium to implement modern portfolio theory since the TSE has remained the focal point for investors looking to invest in Asia's largest stock market, ranking third in the world behind the New York Stock Exchange and NASDAQ in 2018 [2]. Japan has also been recognized as a member of the G7, which includes the world's six major advanced economies: Canada, France, Italy, the United Kingdom, the United States, and Germany. However, in the last decade, Japan has undergone an economic crisis period and the emergence of China has become a threat to Japan's economy. At some point, the systematic risk that occurred might affect the investor's portfolio thus leaving the greatest risk if there is no proper plan in constructing the best portfolio strategy.

Hence, the mean-variance portfolio optimization remains widely used among investors in analyzing their portfolio investment due to its simplicity and ease of derivation [3]. Therefore, there are many previous researchers had done their study using this modern portfolio theory that was first introduced by Harry Markowitz. Kulali [1] claimed that individuals are mainly aiming to select the best diversification through these two related strategies as maximizing return or minimizing risk despite given the characteristics of the country's stock market.

So, is the optimal portfolio being always the best portfolio diversification strategy? A lot of researchers have a debate regarding the issues of the mean-variance approach versus the 1/N strategy. Few researchers such as [4–7] conducted to determine the best portfolio strategy on the various samples of the equity market. Similar results are found that naïve diversification has performed better compared to the optimization model. Ramilton [8], on the other hand, refuted DeMiguel et al. [6]'s claim that a naive portfolio is preferable to an ex-ante portfolio was preferable over the ex-ante optimal portfolio. He found out that the optimal portfolios significantly outperform the naïve portfolio strategy in terms of the Sharpe ratio. Other findings such as [9] explained that if the naïve model outperforms a more sophisticated model, it is a clear indicator that the modeling of the data generating process is not accurate enough. Other research [10–12] used the mean-variance model to optimize asset allocation, therefore, the result showed the mean-variance model outperformed the naïve diversification strategy.

The aim of this paper is briefly to construct the optimal portfolio by using by Markowitz mean-variance model and compare to the naïve diversification strategy in the context of large-cap stock in Tokyo Stock Exchange (TSE) over the three subperiods which are during the global financial crisis (GFC) (2008–2010), the post-global financial crisis (2011–2013), and the non-crisis period (2014–2018).

2. Materials and methods

This study mainly focuses on the model parameters in determining the portfolio optimization construction by using mean-variance analysis on the large-cap stocks in Tokyo Stock Exchange (TSE). Then, we determine the best portfolio diversification strategy in the three subperiods tested.

2.1 Data

The historical stock prices of the Japanese stocks are collected from the Bloomberg Terminal web page and treated as a primary source of data for this study. About 571 data of the stock prices of each stock were picked from the last trading day of every week and were then transformed to weekly return (adjusted price for Japanese Yen). The analysis based on the weekly return is conducted to avoid the non-synchronous trading effect [4].

2.1.1 Stock selection

Tokyo Stock Exchange has remained the largest stock exchange in Asia in terms of market capitalization. Thus, **Table 1** shows about 10 large-cap stocks were chosen randomly upon the stocks that listed in the TOPIX Core 30 + 70 Large and are traded in Tokyo Stock Exchange. The selections of the stocks are the companies that already been existed during the period of this study.

2.1.2 Time periods

The construction of the portfolio of the stocks will be based on 10 years. To look for the workability and superiority of the portfolio strategy in a different timeframe, the 10 years will be divided into three subperiods which are during the global financial crisis (GFC) (2008–2010), the post-global financial crisis (2011–2013), and during the non-crisis period (2014–2018).

	Stocks code	Company name	Industry
1	7201.T	Nissan Motor Co. Ltd.	Automobiles and transportation equipment
2	7267.T	Honda Motor Co. Ltd	Automobiles and transportation equipment
3	8411.T	Mizuho Financial Group, Inc.	Bank
4	8306.T	Mitsubishi UFJ Financial Group, Inc.	Bank
5	8035.T	Tokyo Electron	Electric appliances and precision instruments
6	6702.T	Fujitsu	Electric appliances and precision instruments
7	8801.T	Mitsui Fudosan	Real estate
8	8802.T	Mitsubishi Estate	Real estate
9	9342.T	Nippon Telegraph and Telephone Corporation	Information and communication
10	9433.T	KDDI Corporation	Information and communication

Table 1.
Data description for large-cap stock (TOPIX Core 30 + 70 Large).

2.2 Model framework of portfolio optimization

The rate of return determines whether the investors gain or lose money from their investment [12]. In this study, we calculate the weekly stock return for stock *i* at time *t*, as shown in the Eq. (1).

$$ri = \frac{Sit - Sit - 1}{Sit - 1} \tag{1}$$

where S_{it} is the closing price of stock *i* at time *t*. $S_{it} - 1$ be the closing price at time $t - 1$. We assume that stock *i* pays dividends.

We calculate the average return of stocks based on the following equation:

$$\mu i = E(ri) = \frac{1}{M}\sum_{t=1}^{M} rit \tag{2}$$

where is an average return on stock i, is a market return of stocks i at time t, M is the number of weeks. Then, García et al. [13] defines the expected return of a portfolio is the weighted average of the expected returns of individual stocks. Thus, the equation of expected return for n asset is as the following:

$$E(rp) = \sum_{i=1}^{n} wiE(ri) \tag{3}$$

where is the proportion of the funds invested in stock i, n is the number of stocks, and are the return of ith stock and the return of portfolio p, respectively.

Garcia et al. [10] consider risk as uncertainty through the variability of future returns. So, we calculate the variance of stock i on the weekly return and the index return using the historical volatility formula as in Eq. (4). The higher value in variance for an expected return, the higher the dispersion of expected returns, and the greater the risk of the investment [14].

$$\sigma_i^2 = Var(r_i) = \frac{\sum_{i=1}^{n} (r_i - \mu_i)^2}{M - 1} \tag{4}$$

where,
$Var(r_i)$ is a variance of weekly stock return,
r_i is a weekly stock return,
μ_i is an average weekly return,
M is the sample size.

Ivanovic et al. [12] were able to measure how the stocks vary together by determining the covariance of each stock. The dimensions of risk are organized in the covariance matrix which is denoted by $\Omega_{n \times n}$. This matrix contains variance in its main diagonal and covariances between all pairs of stocks.

$$\Omega_{n \times n} = \begin{pmatrix} \sigma_1^2 & \sigma_{12} & \cdots & \sigma_{1n} \\ \sigma_{21} & \sigma_2^2 & \cdots & \sigma_{2n} \\ \vdots & \vdots & \ddots & \vdots \\ \sigma_{n1} & \sigma_{n2} & \cdots & \sigma_n^2 \end{pmatrix} \tag{5}$$

where,

$$\sigma_{ij} = Cov(r_i, r_j) = \frac{\sum_{i=1,j=1}^{n} (r_i - \mu_i)(r_j - \mu_j)}{n - 1} \tag{6}$$

Equation (6) can be interpreted as the sum of the distance for each value and from the mean is divided by the number of observations minus one. Then, the covariance enables us to calculate the correlation coefficient, shown as:

$$\rho = \frac{\sigma_{ij}}{\sigma_i \sigma_j} \tag{7}$$

where σ is the standard deviation of each asset. This measure is useful to determine the degree of portfolio risk [1].

In this study, the standard deviation is used to measure the risk of the portfolio. The standard deviation of the portfolio is the most common statistical indicator of an asset's risk which measures the dispersion around the expected value [12]. By means, the higher the risk, the higher value of standard deviation. Hence, the equation for standard deviation is defined as follows:

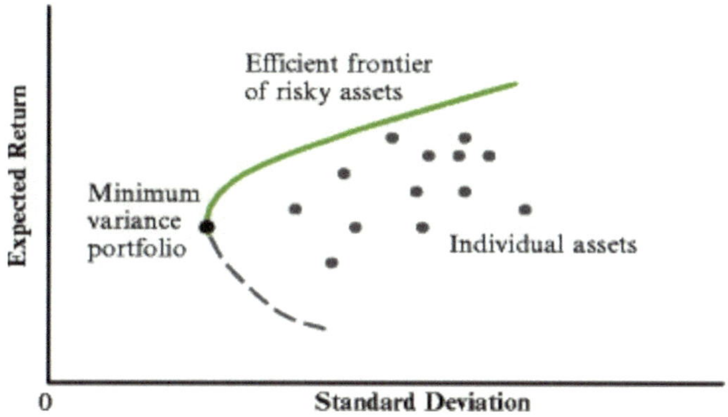

Figure 1.
Efficient frontier [15].

$$\sigma_p = \sqrt{\sum_{i=1}^{n} \left(w_i^2 \cdot \sigma_i^2\right) + 2\left(\sum_{i=1}^{n} \sum_{j=1}^{n} w_i \cdot \sigma_i \cdot w_j \cdot \sigma_j \cdot \rho\right)} \tag{8}$$

where,
σ_p is the standard deviation of a portfolio,
σ_i is the standard deviation of stocks,
w_i is the weight of stocks in a portfolio,
ρ is the correlation coefficient between stock i and j.

Based on the parameters used in the Markowitz mean-variance model, we can get different combinations of expected return and risk. Every possible asset combination is known as an attainable set that can be plotted in risk-return space, and the collection of all such possible portfolios defines a region in this space. **Figure 1** shows that the line along the upper edge of this region is known as the efficient frontier. It is defined as the best return of a portfolio with the lowest risk.

2.3 Performance evaluation ratio

In this study, the Sharpe ratio will be used to measure the portfolio performance that provides risk premium per unit of total risk, which is measured by the portfolio's standard deviation of return. The risk premium is defined as the difference between portfolio return and the risk-free rate. It can be expressed in the following equation:

$$S_p = \frac{\mu_p - R}{\sigma_p} \tag{9}$$

where R is the return on the risk-free asset.

3. Results and discussion

In this study, excel functions and data solvers are used for all calculations. The construction of portfolio optimization using the Markowitz mean-variance approach involves the following steps:

Step 1: Determining return and standard deviation of 10 stocks during three subperiods (**Tables 2–4**).

Step 2: Creating correlations matrix (**Tables 5–7**).

Step 3: Creating variance-covariance matrices during three subperiods (**Tables 8–10**).

Step 4: Calculating the volatility and return of the portfolio with equal weight (**Table 11**).

Step 5: Calculating the volatility and return of the portfolio with difference weights (**Tables 12–14** (all three tables in the Appendices).

Step 6: Creating an efficient frontier (**Figure 2A–C**).

Table 2 reflects during GFC from 2008 to 2010, most of the stock were volatile at high risk and yielding negative returns since the bearish market except Honda Motor, Tokyo Electron, and Mitsui Fudosan which annual returns vary from 0.08 to 6.24%. The risks of 10 stocks vary from 30.05 to 51.63% and Mizuho Financial Group, Inc. has the highest risk recorded during the crisis. As seen from **Table 3**, all

Stock	Average weekly return (GFC)	Weekly variance	Weekly standard deviation	Average annual return	Annual variance	Annual standard deviation
7201.T	−0.0005	0.0035	0.0596	−0.0249	0.1845	0.4296
7267.T	0.0012	0.0036	0.0601	0.0624	0.1876	0.4331
8411.T	−0.0052	0.0051	0.0716	−0.2713	0.2666	0.5163
8306.T	−0.0033	0.0040	0.0633	−0.1693	0.2086	0.4567
8035.T	0.0008	0.0041	0.0643	0.0399	0.2152	0.4639
6702.T	−0.0001	0.0031	0.0554	−0.0065	0.1594	0.3993
8801.T	0.0000	0.0044	0.0665	0.0008	0.2301	0.4796
8802.T	−0.0012	0.0042	0.0648	−0.0622	0.2186	0.4676
9342.T	−0.0013	0.0017	0.0417	−0.0684	0.0903	0.3005
9433.T	−0.0023	0.0021	0.0458	−0.1177	0.1092	0.3305

Table 2.
Risk and return of stocks during global financial crisis (GFC) (2008–2010).

Stocks	Average weekly return (post-GFC)	Weekly variance	Weekly standard deviation	Average annual return	Annual variance	Annual standard deviation
7201.T	0.0017	0.0017	0.0414	0.0882	0.0891	0.2986
7267.T	0.0027	0.0017	0.0410	0.1405	0.0875	0.2958
8411.T	0.0031	0.0014	0.0380	0.1603	0.0752	0.2743
8306.T	0.0035	0.0015	0.0393	0.1845	0.0803	0.2834
8035.T	0.0018	0.0023	0.0476	0.0931	0.1177	0.3431
6702.T	0.0009	0.0022	0.0471	0.0453	0.1153	0.3395
8801.T	0.0065	0.0024	0.0490	0.3373	0.1249	0.3533
8802.T	0.0057	0.0023	0.0477	0.2968	0.1181	0.3436
9342.T	0.0030	0.0007	0.0271	0.1552	0.0383	0.1957
9433.T	0.0073	0.0016	0.0404	0.3799	0.0849	0.2914

Table 3.
Risk and return of stocks during post-global financial crisis (2011–2013).

Stocks	Average weekly return (post-GFC)	Weekly variance	Weekly standard deviation	Average annual return	Annual variance	Annual standard deviation
7201.T	0.001	0.001	0.035	0.030	0.062	0.249
7267.T	−0.001	0.002	0.041	−0.049	0.087	0.296
8411.T	−0.001	0.001	0.038	−0.028	0.076	0.275
8306.T	0.000	0.002	0.040	0.000	0.082	0.286
8035.T	0.004	0.002	0.048	0.217	0.119	0.345
6702.T	0.002	0.002	0.048	0.101	0.121	0.349
8801.T	−0.001	0.002	0.050	−0.041	0.129	0.359
8802.T	−0.002	0.002	0.049	−0.079	0.125	0.353
9342.T	0.002	0.001	0.027	0.115	0.039	0.198
9433.T	0.001	0.002	0.040	0.071	0.084	0.290

Table 4.
Risk and return of stocks during non-crisis period (2014–2018).

	7201.T	7267.T	8411.T	8306.T	8035.T	6702.T	8801.T	8802.T	9342.T	9433.T
7201.T	1.0000									
7267.T	0.7219	1.0000								
8411.T	0.2307	0.3298	1.0000							
8306.T	0.4480	0.5110	0.4336	1.0000						
8035.T	0.6080	0.5780	0.3285	0.5550	1.0000					
6702.T	0.4944	0.5349	0.3392	0.4581	0.5457	1.0000				
8801.T	0.4380	0.4951	0.2525	0.6559	0.5903	0.3973	1.0000			
8802.T	0.4247	0.4908	0.2504	0.6595	0.6089	0.4247	0.9345	1.0000		
9342.T	0.1628	0.1932	0.6257	−0.0272	0.1526	0.0968	0.0145	0.0151	1.0000	
9433.T	0.0116	0.1050	0.3155	0.0058	0.1032	0.1025	−0.0161	−0.0193	0.5407	1.0000

Table 5.
Correlation matrix during GFC (2008–2010).

stocks are yielding positive returns ranging from 4.53 to 33.73% during the period of post-GFC. Mitsui Fudosan lead the market with the highest annual return and the lowest annual return is Fujitsu. Despite the fact that the annual returns show a difference, both stocks are significantly risky, with risk levels ranging between 35.33 and 33.95%. **Table 4** shows during the non-crisis period, about four stocks have poor performance with a negative rate of return. The remaining stocks are ranging from 0.00 to 21.7%. Tokyo Electron is dominating the market with the highest return and the Mitsubishi UFJ Financial Group, Inc. is recovering with the lowest return. The risks of each stock are varying from 19.8 to 35.3% which is dominated by the Mitsui Fudosan. The lowest risk recorded is from Nippon Telegraph and Telephone Corporation. Thus, apart from the results of the risk and return of the individual stock from the three subperiods, the variation of risks per weekly and annual basis has indicated that there is a chance of high variability of investment.

	7201.T	7267.T	8411.T	8306.T	8035.T	6702.T	8801.T	8802.T	9342.T	9433.T
7201.T	1.0000									
7267.T	0.7637	1.0000								
8411.T	0.5127	0.6103	1.0000							
8306.T	0.5722	0.6906	0.8291	1.0000						
8035.T	0.5211	0.5929	0.4321	0.4715	1.0000					
6702.T	0.5485	0.6442	0.5812	0.5465	0.4767	1.0000				
8801.T	0.4850	0.5510	0.5963	0.7078	0.4016	0.4785	1.0000			
8802.T	0.4617	0.5707	0.5644	0.6776	0.3914	0.4632	0.9088	1.0000		
9342.T	0.4512	0.4152	0.5442	0.5402	0.4066	0.4524	0.3644	0.2784	1.0000	
9433.T	0.0571	0.1661	0.1578	0.1544	−0.0024	0.0445	0.1061	0.1138	−0.0014	1.0000

Table 6.
Correlation matrix during post-GFC (2010–2013).

	7201.T	7267.T	8411.T	8306.T	8035.T	6702.T	8801.T	8802.T	9342.T	9433.T
7201.T	1.0000									
7267.T	0.7103	1.0000								
8411.T	0.7115	0.7206	1.0000							
8306.T	0.6906	0.7406	0.9125	1.0000						
8035.T	0.4873	0.4708	0.4467	0.4460	1.0000					
6702.T	0.3999	0.5016	0.4878	0.5004	0.4169	1.0000				
8801.T	0.5912	0.6267	0.6363	0.6740	0.3856	0.3891	1.0000			
8802.T	0.5492	0.6149	0.6176	0.6445	0.3753	0.3685	0.8805	1.0000		
9342.T	0.3704	0.3668	0.3275	0.3107	0.2708	0.1538	0.4520	0.4683	1.0000	
9433.T	0.3592	0.3631	0.3555	0.3454	0.2831	0.1806	0.4431	0.5035	0.6966	1.0000

Table 7.
Correlation matrix during the non-crisis period (2014–2018).

Stock	7201.T	7267.T	8411.T	8306.T	8035.T	6702.T	8801.T	8802.T	9342.T	9433.T
7201.T	0.0035	0.0026	0.0016	0.0017	0.0023	0.0016	0.0017	0.0016	0.0004	0.0007
7267.T	0.0026	0.0036	0.0020	0.0019	0.0022	0.0018	0.0020	0.0019	0.0005	0.0009
8411.T	0.0016	0.0020	0.0051	0.0039	0.0024	0.0020	0.0029	0.0027	0.0010	0.0008
8306.T	0.0017	0.0019	0.0039	0.0040	0.0023	0.0016	0.0027	0.0027	0.0008	0.0009
8035.T	0.0023	0.0022	0.0024	0.0023	0.0041	0.0019	0.0025	0.0025	0.0005	0.0010
6702.T	0.0016	0.0018	0.0020	0.0016	0.0019	0.0031	0.0014	0.0015	0.0009	0.0008
8801.T	0.0017	0.0020	0.0029	0.0027	0.0025	0.0014	0.0044	0.0041	0.0008	0.0014
8802.T	0.0016	0.0019	0.0027	0.0027	0.0025	0.0015	0.0041	0.0042	0.0008	0.0012
9342.T	0.0004	0.0005	0.0010	0.0008	0.0005	0.0009	0.0008	0.0008	0.0017	0.0012
9433.T	0.0007	0.0009	0.0008	0.0009	0.0010	0.0008	0.0014	0.0012	0.0012	0.0021

Table 8.
Variance-covariance matrix during GFC.

Stock	7201.T	7267.T	8411.T	8306.T	8035.T	6702.T	8801.T	8802.T	9342.T	9433.T
7201.T	0.0017	0.0013	0.0008	0.0009	0.0010	0.0011	0.0010	0.0009	0.0005	0.0003
7267.T	0.0013	0.0017	0.0010	0.0011	0.0012	0.0013	0.0011	0.0011	0.0005	0.0004
8411.T	0.0008	0.0010	0.0015	0.0012	0.0008	0.0010	0.0011	0.0010	0.0006	0.0005
8306.T	0.0009	0.0011	0.0012	0.0016	0.0009	0.0010	0.0014	0.0013	0.0006	0.0005
8035.T	0.0010	0.0012	0.0008	0.0009	0.0023	0.0011	0.0009	0.0009	0.0005	0.0006
6702.T	0.0011	0.0013	0.0010	0.0010	0.0011	0.0022	0.0011	0.0011	0.0006	0.0004
8801.T	0.0010	0.0011	0.0011	0.0014	0.0009	0.0011	0.0024	0.0021	0.0005	0.0005
8802.T	0.0009	0.0011	0.0010	0.0013	0.0009	0.0011	0.0021	0.0023	0.0003	0.0005
9342.T	0.0005	0.0005	0.0006	0.0006	0.0005	0.0006	0.0005	0.0003	0.0007	0.0005
9433.T	0.0003	0.0004	0.0005	0.0005	0.0006	0.0004	0.0005	0.0005	0.0005	0.0016

Table 9.
Variance-covariance matrix during post-GFC.

Stock	7201.T	7267.T	8411.T	8306.T	8035.T	6702.T	8801.T	8802.T	9342.T	9433.T
7201.T	0.0020	0.0014	0.0014	0.0018	0.0014	0.0011	0.0014	0.0012	0.0006	0.0007
7267.T	0.0014	0.0021	0.0014	0.0020	0.0013	0.0014	0.0016	0.0015	0.0006	0.0007
8411.T	0.0014	0.0014	0.0020	0.0023	0.0012	0.0013	0.0015	0.0014	0.0005	0.0007
8306.T	0.0018	0.0020	0.0023	0.0034	0.0016	0.0017	0.0021	0.0019	0.0007	0.0009
8035.T	0.0014	0.0013	0.0012	0.0016	0.0040	0.0016	0.0013	0.0012	0.0006	0.0008
6702.T	0.0011	0.0014	0.0013	0.0017	0.0016	0.0035	0.0013	0.0011	0.0004	0.0005
8801.T	0.0014	0.0016	0.0015	0.0021	0.0013	0.0013	0.0030	0.0025	0.0010	0.0011
8802.T	0.0012	0.0015	0.0014	0.0019	0.0012	0.0011	0.0025	0.0027	0.0009	0.0012
9342.T	0.0006	0.0006	0.0005	0.0007	0.0006	0.0004	0.0010	0.0009	0.0015	0.0012
9433.T	0.0007	0.0007	0.0007	0.0009	0.0008	0.0005	0.0011	0.0012	0.0012	0.0021

Table 10.
Variance-covariance matrix during non-crisis period.

Period	Portfolio return	Portfolio variance	Portfolio standard deviation
During GFC	−0.0012	0.0019	0.0435
Post-GFC	0.0036	0.0010	0.0310
Non-crisis	0.0006	0.0014	0.0372

Table 11.
Portfolio with equal weight (naïve diversification).

Correlation between assets determines the degree of portfolio risk [11]. Negative or small correlation between assets, the risk of the portfolio is low whereas the positive or large correlation between the assets, the risk of the portfolio is high. As seen in **Table 5**, only 9342.T–8306.T, 9433.T–8801.T, and 9433.T–8802.T have a negative correlation between the stocks and others are greater than zero. **Table 6** reflects that only two pairs of stocks which are 9433.T–8035.T and 9433.T–9342.T have a negative correlation and **Table 7** has all positive correlation between the stocks and not low enough. In that sense, only during the non-crisis period the portfolio constructed will not be well diversified.

When there are more than two assets, covariance can best be calculated by using matrix algebra based on Eq. (6). The covariance calculation which involved the excess return of stocks and the number of observations in each subperiod allows us to compute by using few functions in excel such as @MMULT(...) and TRANSPOSE(...). The procedure has already allowed us to determine the variance-covariance matrix as per **Tables 8–10**. The values of the variance-covariance matrices have shown that in the three subperiods tested, all the stocks are move in the same direction with a lower degree. Thus, this indicates that the volatility can be reduced during the construction of the portfolios.

The variance-covariance matrix helps in a simple way of measuring portfolio variance. At this step, the portfolio return and portfolio variance can be calculated. In this sense, the naïve allocation of weight whereby the equal proportion in the portfolio is used to calculate the portfolio return and portfolio variance. Since we have 10 stocks, each stock will have about 1/10 or 10% weight.

So, **Table 11** shows that the portfolio returns and portfolio variance of the naïve diversification strategy. Among the three subperiods, during GFC, the portfolio indicates poor performance by having a negative return of −0.12% and the highest standard deviation of 4.35%. In contrast, the portfolio return is recorded high during post-GFC which is about 0.36% with the lowest standard deviation of 3.10%.

Note that, Markowitz's mean-variance model stated that the investor must be rational and risk-averse [16]. Thus, in constructing the efficient portfolios, two conditions need to be satisfied which are maximum return for varying levels of risk and minimum risk for varying levels of expected return. In the context of three subperiods tested, this study proposed rational investors for looking at optimal portfolio with a minimum level of risk. Thus, this optimal portfolio analysis can be shown as the subject function according to the formula:

$$Min\sigma^2 \sum_{j=1}^{n} w_i w_j Cov_{ij} \tag{10}$$

where w_i and w_j are weights of stocks in the portfolio and Cov_{ij} is the covariance value between stock i and j.

Then, three main constrains in Markowitz mean-variance portfolio optimization are included for the optimization problem. The formulas are written as:

$$\sum_{i=1}^{n} w_i E(r_i)^3 \geq E* \tag{11}$$

$$\sum_{i=1}^{n} w_i = 1 \tag{12}$$

and

$$wi \geq 0; i = 1, \ldots, N \tag{13}$$

where $E(R_i)$ is the target expected return, E^* is an expected return and w_i is the weight of the stock i. The third constraint is added to restrict the short sell to happen.

Hence, Excel Solver is used to optimizing the weight by including all the constraints according to Eqs. (10)–(13). About 10 iterations were done in Solver to

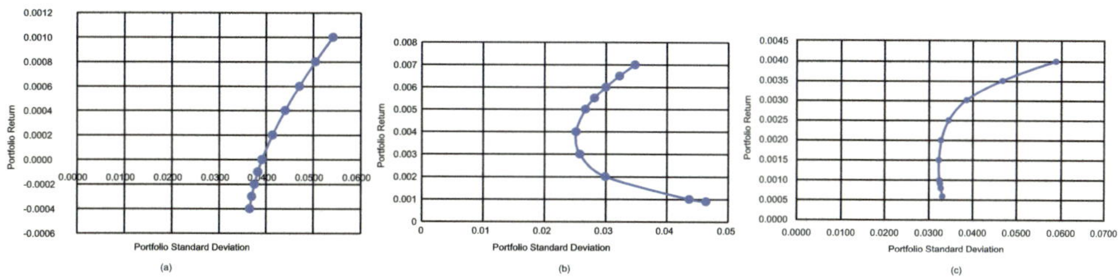

Figure 2.
(a) Efficient frontier of portfolios during GFC, (b) efficient frontier of portfolios during post-GFC, and (c) efficient frontier of portfolios during the non-crisis period.

generate 10 portfolios. The 10 portfolios that are generated produce 10 different sets of weights for the level of return which minimizes the standard deviation of the portfolio. All the results are shown in **Tables 12–14** (in Appendices) indicate the composition of stocks, portfolio return, and portfolio risk. All the set of possible expected return and risk combinations is then called attainable set [1]. The attainable set is plotted on the risk-return space as shown in **Figure 2A–C**.

Figure 2A–C demonstrate the efficient frontier with minimum variance portfolios. Ivanova et al. [11] states that given expected return, investors could choose an optimal portfolio based on risk preference since all the efficient portfolios are located over the efficient frontier curve. Many portfolios are existed on the efficient frontier curve with different risk and return combinations. Then, the optimal portfolio is selected on the efficient frontier curve in each subperiods then compare with the portfolio return and risk for naïve diversification strategy.

Table 15 reflects the analysis on the risk, return and performance of the portfolio constructed by using naïve diversification strategy and Markowitz mean-variance portfolio optimization. In the three subperiods tested, the optimal portfolio outperformed the naïve diversification strategy.

In terms of portfolio risk-return, optimal portfolios selected during global financial crisis tend to have a high standard deviation of 4.71% compared to naïve diversification which is about 4.35%, respectively. But, the portfolio with equal weight yields an abnormal return of −0.12% compared to the optimal portfolio of 0.06%. This is not a surprising result since the turbulence of economic recession contributes to the high-risk investment. During the post-GFC, again the optimal portfolio yields a higher return with 0.5% with lower risk at 2.67% compared to the naïve strategy. Furthermore, there is not much difference in standard deviation for both strategies in the non-crisis period, but the optimal portfolio still dominates with a high return at 2.5% and low risk at 3.44%.

Sharpe ratio classifies during financial crisis both strategies produce negative Sharpe ratio which indicates the portfolio's return is less than the risk-free rate. Thus, the interpretation of the negative Sharpe ratio will be put aside since it does not convey any useful meaning. For the rest two periods, the Sharpe ratio is undoubtedly the highest for the optimal portfolio compare to the naïve portfolio. This shows the reward-to-volatility ratio when investing in the optimal portfolio 6 and portfolio 7 are 0.1032 and 0.0582.

4. Conclusion

This study was aimed to help investors to plan for the best investment strategy in maximizing return with the given level of risk or minimizing risk. In this study, there are subperiods been tested for this whole study, during the global financial crisis (GFC) 2008–2010, the post-financial crisis 2011–2013 and non-financial crisis 2014–2018. Then, we followed the Markowitz mean-variance model which involved the best possible combination of expected return and risk to construct efficient portfolios in each period. The portfolio which did not contain short sale was achieved by a data solver. We made a choice to choose the optimal portfolio from this efficient set based on rational investors.

Meantime, we also constructed a portfolio with equal weight as a control parameter to determine the best diversification strategy. It was figured out during the period of study; the optimal portfolio was chosen from the efficient frontier formed by Markowitz model perform better than the portfolio with equal weight in all three subperiods. The result reiterates the previous study conducted by [1, 9, 11]. Hence, the Markowitz framework is successful since the variance-covariance and correlation played an important role in determining the optimal portfolio. So, if the Japanese and foreign investors know properly how to apply the Markowitz mean-variance model in their investment. We believe this is the best solution in many alternatives.

But, the limitation of this study is there is no out-of-sample data tested. For future research, a robust optimization approach can be considered with the out-of-sample data tested to construct the optimal portfolio. The portfolio also needs to be rebalanced again to ensure a more accurate result.

Acknowledgements

I would to express my deepest appreciation to all those who provided me the possibility to complete this research paper. A special gratitude I give to my supervisor Dr. Norizarina binti Ishak and my friend Nur Fathin Shaida Binti Muhammad Nadhirin whose contribution in stimulating suggestions and encouragement, helped me to coordinate my research especially in writing this research paper.

Appendices

	Portfolio return	Portfolio standard deviation	Weight of stock									
			7201.T	7267.T	8411.T	8306.T	8035.T	6702.T	8801.T	8802.T	9342.T	9433.T
Portfolio 1	−0.0004	0.0366	0.0441	0.2204	0	0	0.1213	0.0578	0.0071	0	0.5418	0.0075
Portfolio 2	−0.0003	0.0370	0.0158	0.2558	0	0	0.1360	0.0576	0.0050	0	0.5298	0
Portfolio 3	−0.0002	0.0376	0	0.2927	0	0	0.1503	0.0552	0.0002	0	0.5017	0
Portfolio 4	−0.0001	0.0383	0	0.3268	0	0	0.1602	0.0500	0	0	0.4630	0
Portfolio 5	0.0000	0.0392	0	0.3607	0	0	0.1701	0.0450	0	0	0.4242	0
Portfolio 6	0.0002	0.0413	0	0.4285	0	0	0.1899	0.0351	0	0	0.3465	0
Portfolio 7	0.0004	0.0440	0	0.4964	0	0	0.2097	0.0252	0	0	0.2688	0
Portfolio 8	0.0006	0.0471	0	0.5642	0	0	0.2294	0.0152	0	0	0.1911	0
Portfolio 9	0.0008	0.0505	0	0.6321	0	0	0.2492	0.0053	0	0	0.1134	0
Portfolio 10	0.0010	0.0541	0	0.6988	0	0	0.2676	0	0	0	0.0335	0

Table 12.
Composition of weight and risk of different expected return portfolios during GFC period.

	Portfolio return	Portfolio standard deviation	Weight of stock									
			7201.T	7267.T	8411.T	8306.T	8035.T	6702.T	8801.T	8802.T	9342.T	9433.T
Portfolio 1	0.0009	0.0464	0.0350	0	0	0	0	0.9650	0	0	0	0
Portfolio 2	0.0010	0.0437	0.1561	0	0	0	0	0.8439	0	0	0	0
Portfolio 3	0.0020	0.0299	0.2235	0	0	0	0.0862	0.2812	0	0	0.4091	0
Portfolio 4	0.0030	0.02588	0.1318	0	0.0182	0	0.0377	0.0341	0	0.0425	0.6933	0.0426
Portfolio 5	0.0040	0.0252	0.0710	0	0	0	0	0	0	0.1197	0.6170	0.1813
Portfolio 6	0.0050	0.0267	0	0	0	0	0	0	0.0743	0.1165	0.4765	0.3327
Portfolio 7	0.0055	0.0281	0	0	0	0	0	0	0.1656	0.0522	0.3673	0.4150
Portfolio 8	0.0060	0.03	0	0	0	0	0	0	0.2467	0	0.2555	0.4978
Portfolio 9	0.0065	0.0322	0	0	0	0	0	0	0.2842	0	0.1326	0.5832
Portfolio 10	0.0070	0.0347	0	0	0	0	0	0	0.3216	0	0.0098	0.6686

Table 13.
Composition of weight and risk of different expected return portfolios during post-GFC period.

	Portfolio return	Portfolio standard deviation	Weight of stock									
			7201.T	7267.T	8411.T	8306.T	8035.T	6702.T	8801.T	8802.T	9342.T	9433.T
Portfolio 1	0.0006	0.0331	0.0579	0.1859	0.2419	0	0	0.0389	0	0.0465	0.3273	0.1016
Portfolio 2	0.0008	0.0327	0.0774	0.1554	0.2273	0	0	0.0585	0	0.0195	0.3602	0.1018
Portfolio 3	0.0009	0.0325	0.0869	0.1401	0.2198	0	0	0.0683	0	0.0062	0.3767	0.1020
Portfolio 4	0.0010	0.0323	0.0985	0.1192	0.2085	0	0	0.0795	0	0	0.3946	0.0997
Portfolio 5	0.0015	0.0322	0.1379	0.0138	0.1478	0	0.0305	0.1212	0	0	0.4687	0.0801
Portfolio 6	0.0020	0.0328	0.1324	0	0.0393	0	0.0969	0.1442	0	0	0.5373	0.0499
Portfolio 7	0.0025	0.0344	0.0331	0	0	0	0.1938	0.1506	0	0	0.6226	0
Portfolio 8	0.0030	0.0385	0	0	0	0	0.4072	0.0402	0	0	0.5525	0
Portfolio 9	0.0035	0.0468	0	0	0	0	0.6586	0	0	0	0.3414	0
Portfolio 10	0.0040	0.0587	0	0	0	0	0.9157	0	0	0	0.0843	0

Table 14.
Composition of weight and risk of different expected return portfolios during non-crisis period.

Period	Diversification strategy	Portfolio return	Portfolio standard deviation	Sharpe ratio
During GFC (2008–2009)	Naïve strategy	−0.0012	0.0435	−0.1047
	Optimal portfolio 8	0.0006	0.0471	−0.0349
Post-GFC (2011–2013)	Naïve strategy	0.0036	0.031	0.0437
	Optimal Portfolio 6	0.005	0.0267	0.1032
Non-crisis period (2014–2018)	Naïve strategy	0.0006	0.0372	0.0027
	Optimal portfolio 7	0.0025	0.0344	0.0582

Table 15.
Comparison of portfolio risk, return, and performance evaluation.

Author details

Muhammad Jaffar Sadiq Abdullah and Norizarina Ishak*
Faculty of Science and Technology, Universiti Sains Islam Malaysia, Nilai, Malaysia

*Address all correspondence to: norizarina@usim.edu.my

References

[1] Kulali I. Portfolio optimization analysis with Markowitz quadratic mean-variance model. European Journal of Business and Management. 2016; **8**(7):73-79

[2] Shukla V. Top 10 Largest Stock Exchanges in the World by Market Capitalization. Valuewalk [Internet]. 2019. Available from: https://www.valuewalk.com/2019/02/top-10-largest-stock-exchanges/

[3] Shalit H, Yitzhaki S. The mean-Gini efficient portfolio frontier. Journal of Financial Research. 2005; **28**(1):59-75

[4] Baumöhl E, Lyócsa Š. Constructing weekly returns based on daily stock market data: A puzzle for empirical research? In: MPRA Paper 43431. Germany: University Library of Munich; 2012

[5] Brown SJ, Hwang I, In F. Why optimal diversification cannot outperform naive diversification: Evidence from tail risk exposure. SSRN Electronic Journal. 2013:1-55. DOI: 10.2139/ssrn.2242694

[6] DeMiguel V, Garlappi L, Uppal R. Optimal versus naive diversification: How inefficient is the 1/N portfolio strategy? Review of Financial Studies. 2009;**22**(5):1915-1953

[7] Gupta M, Aggarwal N. Naïve versus mean-variance diversification in Indian capital markets. Asia-Pacific Journal of Management Research and Innovation. 2015;**11**(3):198-204

[8] Ramilton A. Should you optimize your portfolio?: On portfolio optimization: The optimized strategy versus the naïve and market strategy on the Swedish stock market. 2014. Available from: http://urn.kb.se/resolve?urn=urn:nbn:se:uu:diva-218024

[9] Pflug GC, Pichler A, Wozabal D. The 1/N investment strategy is optimal under high model ambiguity. Journal of Banking & Finance. 2012;**36**(2):410-417. DOI: 10.1016/j.jbankfin.2011.07.018

[10] Garcia T, Borrego D. Markowitz efficient frontier and capital market line—Evidence from the Portuguese. Portuguese Journal of Management Studies. 2017;**22**(1):3-23

[11] Ivanova M, Dospatliev L. Application of Markowitz portfolio optimization on Bulgarian stock market from 2013 to 2016. International Journal of Pure and Applied Mathematics. 2017;**117**(2): 291-307. DOI: 10.12732/ijpam.v117i2.5

[12] Ivanovic Z, Baresa S, Bogdan S. Portfolio optimization on Croatian capital market. UTMS Journal of Economics. 2013;**4**(3):269-282

[13] García F, González-Bueno JA, Oliver J. Mean-variance investment strategy applied in emerging financial markets: Evidence from the Colombian stock market. Intellectual Economics. 2015;**9**(1):22-29

[14] Sun Y. Optimization stock portfolio with mean-variance and linear programming: Case in Indonesia stock market. Binus Business Review. 2010; **1**(1):15. DOI: 10.21512/bbr.v1i1.1018

[15] Chen WP, Chung H, Ho KY, Hsu TL. Portfolio optimization models and mean–variance spanning tests. In Handbook of quantitative finance and risk management. Boston, MA: Springer. 2010;165-184

[16] Markowitz H. Portfolio Selection*. The Journal of Finance. 1952;7:77-91. DOI: 10.1111/j.1540-6261.1952.tb01525.x

Chapter 2

Existence, Uniqueness and Stability of Fractional Order Stochastic Delay System

Sathiyaraj Thambiayya, P. Balasubramaniam, K. Ratnavelu and JinRong Wang

Abstract

This chapter deals with the problem of fractional higher-order stochastic delay systems. A solution representation is given by using sin and cos matrix functions for different delay intervals. Further, existence and uniqueness results are proved through fixed point theorem. Moreover, finite-time stability criteria are obtained using fractional Gronwall-Bellman inequality lemma. Finally, numerical simulation is carried out to check the proposed theoretical results.

Keywords: existence and uniqueness of solution, fixed point theorem, fractional differential equations (FDEs), stochastic differential system

1. Introduction

Fractional derivatives (FD) initiative concept is quite old and its history spans three centuries. The variety of papers dedicated to FD is multiplied swiftly in the mid-twentieth century and later decades. One of the motives for the full-size interest within the discipline of FD is that it's far feasible to describe the variety of physical [1], synthetic [2], and organic [3] occurrence with fractional differential equations (FDEs). As a new branch of applied mathematics, the field of FD can be seen in many applications. Nevertheless, more and more compelling implementations have been found in various engineering and science fields over the past few decades (see [4]). It is noted that the existing theory of FDEs is committed to a larger part of the research works (see [5–8]). While modeling functional structures, ambient noise and time delays need to be taken into account, which might be very beneficial in building extra sensible fashions of sciences, and so on [9]. It is referred to as the pattern direction houses of the stochastic fractional partial differential gadget powered by way of area time white noise [10].

The problems in a stochastic environment replicate the modeling of single-sever m-mode random queues in computer networks [11], the spatial distribution of mobile users in the telecommunications network coverage area [12], and other anomalies that occurred naturally in many disciplines [13]. Authors in [14] investigated the existence, uniqueness, and large deviation principle solutions to stochastic evolution equations of jump type. Among the many meaningful properties of stochastic stability results describe the maximum vital feature of fractional order stochastic systems and have been investigated in Refs. [15–18]. The notion of finite-time stability for fractional stochastic delay systems occurs a matter of course in stochastic control systems. Without any doubt that this type of fractional stochastic stability is most important in both theory and applications.

However, only few introductions and discussions exist on the definition of finite-time stability in stochastic finite space using fixed point theorem approach. Burton [19] started to analyze the stability characters of dynamical systems broadly using fixed point theorems. Subsequently, few authors applied fixed point approach to establish sufficient conditions for stability of the differential systems (see [20–24]). Based on the above discussions, this chapter provides finite-time stability of the Caputo sense FDEs via fixed point theorems.

The primary contribution of this chapter is defined as follows:

i. A fractional higher-order stochastic delay system (FSDS) is considered in finite-dimensional stochastic settings.

ii. Weaker hypothesis on nonlinear terms and appropriate fixed-point analysis are utilized to obtain the existence and uniqueness of solution.

iii. A new set of generalized sufficient conditions for finite-time stability of a certain FSDS is established by using Generalized Gronwall-Bellman inequality.

Novelties and challenges of this chapter are described through the subsequent statements:

i. Finite-time stability analysis for FSDS is new in literature of finite-dimensional fractional stochastic settings.

ii. It is a challenge to tackle the proposed system with a norm estimation on nonlinear stochastic terms as described in this chapter.

iii. It is more complex to verify the weaker assumptions of the system and the derived result is new, has not been analyzed with the existing literature.

iv. Obtained result is proved in stochastic nature with square norm settings.

Organization of this chapter is as follows: system description and preliminaries are provided in Section 2. Existence and uniqueness of solution are provided in Section 3. Finite-time stability result is proved in Section 4 and Section 5 consists of a numerical example.

Notations: ${}^{C}D^{q}_{-\kappa^+}$ represent respectively the Caputo derivative with $q \in (0,1)$; R^n and $R^{n\times n}$ represent the $n-$dimensional Euclidean space and $n \times n$ real matrix; $E(\cdot)$ denotes the mathematical expectation with some probability measure; $\Omega = \left(L^2_{F_0}([0,b], R^n), \|\cdot\|\right)$; for any $y \in R^n$, we define the norm

$$\|y(\eta)\| = \sqrt{\sup_{\eta\in[-\kappa,b]} \left\{e^{-2N\eta}|y(\eta)|^2\right\}};$$

define a column-wise matrix sum

$$\|M\| = \max\left\{\sum_{k=1}^{j}|m_{k1}|, \sum_{k=1}^{j}|m_{k2}|, \ldots, \sum_{k=1}^{j}|m_{kjn}|\right\}.$$

Further, let as define the matrix norm

$$\max_{\eta \in [-\kappa, 0]} \left\{ e^{-2N\eta} |\psi(\eta)|^2 \right\} = \|\psi\|^2, \quad \max_{\eta \in [-\kappa, 0]} \left\{ e^{-2N\eta} |\psi'(\eta)|^2 \right\} = \|\psi'\|^2.$$

2. System description and preliminaries

Consider the following system:

$$\begin{cases} {}^C D^q_{-\kappa^+} \left({}^C D^q_{-\kappa^+} y \right)(\eta) + M^2 y(\eta - \kappa) = F(\eta, y(\eta)) + \int_0^\eta \Delta(\zeta, y(\zeta)) dw(\zeta), & \eta \in [0, b], \quad \kappa > 0, \\ y(\eta) = \psi(\eta), \quad y'(\eta) = \psi'(\eta), \quad \eta \in [-\kappa, 0], \quad \kappa > 0, \end{cases} \tag{1}$$

where $y(\eta) \in R^n$ is a state vector. Here, $M \in R^{n \times n}$ is taken as a nonsingular matrix. F is mapping from $[0, b] \times R^n$ to R^n and Δ is a mapping from $[0, b] \times R^n$ to $R^{n \times d}$ are nonlinear continuous function and $\psi \in C^1([-\kappa, 0], R^n)$ is an initial value function. w denotes $d-$dimensional Wiener process.

Definition 2.1. ([5]) *The Caputo derivative for $f : [-\kappa, \infty) \to R$, is*

$$\left({}^C D^q_{-\kappa^+} f \right)(\eta) = \frac{1}{\Gamma(1-q)} \int_{-\kappa}^{\eta} (\eta - \zeta)^{-q} f'(\zeta) d\zeta, \quad q \in (0, 1], \quad \eta > -\kappa, \quad f'(\eta) = \frac{df}{d\eta}.$$

Definition 2.2. *(see [5]) Mittag-Leffler function is*

$$E_{q,p}(u) = \sum_{k=0}^{\infty} \frac{u^k}{\Gamma(kq + p)} \quad \text{for } q, p > 0.$$

In particular, for $p = 1$,

$$E_{q,1}(\theta u^q) = E_q(\theta u^q) = \sum_{k=0}^{\infty} \frac{\theta^k u^{kq}}{\Gamma(qk + 1)}, \quad \theta, u \in \mathbb{C}.$$

Definition 2.3. *(see [25]) The 2kq degree of polynomial for delayed fractional* cos *matrix is given at $\eta = k\kappa$, $k = 0, 1, \cdots$*

$$\cos_{\kappa,q} M\eta^q = \begin{cases} \Theta, & -\infty < \eta < -\kappa, \\ I, & -\kappa \le \eta < 0, \\ \vdots & \vdots \\ I - M^2 \dfrac{\eta^{2q}}{\Gamma(2q+1)} + \cdots\cdots + (-1)^k M^{2k} \dfrac{(\eta - (k-1)\kappa)^{2k\kappa}}{\Gamma(2kq+1)}, & (k-1)\kappa \le \eta < k\kappa, \\ \vdots & \vdots \end{cases}$$

where Θ and I represent the zero and identity matrices.

Definition 2.4. *([25]) The $(2k+1)q$ degree of polynomial for a delayed fractional* sin *matrix is given at $\eta = k\kappa$, $k = 0, 1, \cdots$*

$$\sin_{\kappa,q}M\eta^q = \begin{cases} \Theta, & -\infty < \eta < -\kappa, \\ M\dfrac{(\eta+\kappa)^q}{\Gamma(q+1)}, & -\kappa \le \eta < 0, \\ \vdots & \vdots \\ M\dfrac{(\eta+\kappa)^q}{\Gamma(q+1)} + \cdots\cdots + (-1)^k M^{2k+1}\dfrac{(\eta-(k-1)\kappa)^{(2k+1)q}}{\Gamma[(2k+1)q+1]}, & (k-1)\kappa \le \eta < k\kappa. \\ \vdots & \vdots \end{cases}$$

We have the following square norm estimations:

i. $\| \cos_{\kappa,q}M\eta^q \|^2 = \left(\sum_{k=0}^{\infty} \frac{(\|M\|\eta^{2q})^k}{\Gamma(2kq+1)} \right)^2 \le \left[\mathbb{E}_{2q}\left(\|M\|^2\eta^{2q}\right)\right]^2, \ \eta \in [(k-1)\kappa, k\kappa),$
$k = 0, 1, 2, \cdots n$

ii.
$\| \sin_{\kappa,q}M\eta^q \|^2$

$$= \sum_{k=0}^{\infty} \frac{(\|M\|(\eta+\kappa)^q)^{2k}}{(\Gamma(kq+1))^2} + \sum_{k=0}^{\infty} \frac{\left(\|M\|^2(\eta+\kappa)^{2q}\right)^{2k}}{(\Gamma(2kq+1))^2}$$

$$+2\sum_{k_1=0}^{\infty}\sum_{k_2=0}^{\infty} \frac{(\|M\|(\eta+\kappa)^q)^{k_1}}{\Gamma(k_1q+1)} \frac{\left(\|M\|^2(\eta+\kappa)^{2q}\right)^{k_2}}{\Gamma(2k_2q+1)}$$

$$\le \left[\mathbb{E}_q(\|M\|(\eta+\kappa)^q)\right]^2 + \left[\mathbb{E}_{2q}\left(\|M\|^2(\eta+\kappa)^{2q}\right)\right]^2$$

$$+2\mathbb{E}_q(\|M\|(\eta+\kappa)^q)\mathbb{E}_{2q}\left(\|M\|^2(\eta+\kappa)^{2q}\right), \ \eta \in [(k-1)\kappa, k\kappa), \ k = 0, 1, 2, \cdots n.$$

Definition 2.5. *System* (1) *satisfying* $y(\eta) \equiv \psi(\eta)$ *and* $y'(\eta) = \psi'(\eta)$ *for* $-\kappa \le \eta \le 0$ *is finite-time stable in mean square with respect to* $\{0, [0,b], \delta, \varepsilon, \kappa\}$, *if and only if* $\delta_1 < \delta \ (\delta > 0)$ *implies* $E\|y(\eta)\|^2 < \varepsilon \ (\varepsilon > 0), \ \forall \ \eta \in [0,b]$ *where* $\delta_1 = \max\left\{\|\psi\|^2, \|\psi'\|^2\right\}$ *denotes the initial time of observation of the system.*

Lemma 2.1. *[26] (Generalized Gronwall-Bellman inequality) Let* $v(\eta), \ b(\eta)$ *be nonnegative and locally integrable on* $0 \le \eta < b$ *and let* $h(\eta)$ *be a nonnegative, nondecreasing continuous function defined on* $0 \le \eta < b, \ h(\eta) \le M$, *and let M be a real constant,* $q > 0$ *with*

$$v(\eta) \le b(\eta) + h(\eta)\int_0^\eta (\eta-\zeta)^{q-1}v(\zeta)d\zeta$$

and then

$$v(\eta) \le b(\eta) + \int_0^\eta \left[\sum_{k=1}^{\infty} \frac{(h(\eta)\Gamma(q))^k}{\Gamma(kq)} (\eta-\zeta)^{kq-1}v(\zeta)d\zeta\right].$$

Moreover, if $b(\eta)$ *is a nondecreasing function on* $[0,b]$. *Then*

$$v(\eta) \le b(\eta)E_{q,1}(h(\eta)\Gamma(q)\eta^q), \ \eta \in [0,b],$$

where $E_{q,1}(\cdot)$ is the one parameter Mittag-Leffler function.

Assumption 1: Let $x, y \in R^n$, then we take

$$\sup_{\eta \in [-\kappa, b]} e^{-2N\eta} |x(\eta) - y(\eta)|^2 = E\|x(\eta) - y(\eta)\|^2.$$

Lemma 2.2. *For a nonsingular matrix M, the solution of the inhomogeneous system is*

$$\begin{cases} {}^C D^q_{-\kappa^+}\left({}^C D^q_{-\kappa^+} y\right)(\eta) + M^2 y(\eta - \kappa) = f(\eta), \quad \eta \in [0, b], \quad \kappa > 0, \\ y(\eta) = \psi(\eta), \\ y'(\eta) = \psi'(\eta), \quad \eta \in [-\kappa, 0], \end{cases} \tag{2}$$

for zero initial value has the below form:

$$y(\eta) = \int_0^\eta \cos_{\kappa,q} M(\eta - \kappa - \zeta)^q f(\zeta) d\zeta, \quad \eta \in [0, b].$$

Proof. Consider

$$y(\eta) = \int_0^\eta \cos_{\kappa,q} M(\eta - \kappa - \zeta)^q C(\zeta) d\zeta$$

where $C(\zeta)$ (unknown) $\zeta \in [0, \eta]$. By applying ${}^C D^q_{-\kappa^+}\left({}^C D^q_{-\kappa^+}\right)$ on both sides of the above equation one can obtain

$$\begin{aligned} {}^C D^q_{-\kappa^+}\left({}^C D^q_{-\kappa^+} y\right)(\eta) &= \left(\cos_{\kappa,q} M\eta^q\right) C(\eta) - M^2 \int_0^\eta \cos_{\kappa,q} M(\eta - 2\kappa - \zeta)^q C(\zeta) d\zeta \\ &= C(\eta) - M^2 \int_0^\eta \cos_{\kappa,q} M(\eta - 2\kappa - \zeta)^q C(\zeta) d\zeta \\ &+ M^2 \int_0^{\eta - \kappa} \cos_{\kappa,q} M(\eta - 2\kappa - \zeta)^q C(\zeta) d\zeta. \end{aligned}$$

Substitute the above expression into (2), one can get

$$C(\eta) - M^2 \int_0^\eta \cos_{\kappa,q} M(\eta - 2\kappa - \zeta)^q C(\zeta) d\zeta + M^2 \int_0^{\eta - \kappa} \cos_{\kappa,q} M(\eta - 2\kappa - \zeta)^q C(\zeta) d\zeta = f(\eta),$$

since $\int_{\eta-\kappa}^\eta \cos_{\kappa,q} M(\eta - 2\kappa - \zeta)^q C(\zeta) d\zeta = 0$. Hence the proof. □

Using [25] and Lemma 2.2, the solution of (1) is

$$\begin{aligned} y(\eta) = &\left(\cos_{\kappa,q} M\eta^q\right)\psi(-\kappa) + M^{-1}\left(\sin_{\kappa,q} M(\eta - \kappa)^q\right)\psi'(0) + \int_{-\kappa}^0 \cos_{\kappa,q} M(\eta - \kappa - \zeta)^q \psi'(\zeta) d\zeta \\ &+ \int_0^\eta \cos_{\kappa,q} M(\eta - \kappa - \zeta)^q F(\zeta, y(\zeta)) d\zeta \\ &+ \int_0^\eta \cos_{\kappa,q} M(\eta - \kappa - \zeta)^q \left(\int_0^\zeta \Delta(\lambda, y(\lambda)) dw(\lambda)\right) d\zeta. \end{aligned}$$

Define the nonlinear operator $\mathcal{P} : R^n \to R^n$ by

$$(\mathcal{P}y)(\eta) = \left(\cos_{\kappa,q} M\eta^q\right)\psi(-\kappa) + M^{-1}\left(\sin_{\kappa,q} M(\eta-\kappa)^q\right)\psi'(0)$$
$$+\int_{-\kappa}^{0} \cos_{\kappa,q} M(\eta-\kappa-\zeta)^q \psi'(\zeta)d\zeta + \int_0^\eta \cos_{\kappa,q} M(\eta-\kappa-\zeta)^q F(\zeta,y(\zeta))d\zeta$$
$$+\int_0^\eta \cos_{\kappa,q} M(\eta-\kappa-\zeta)^q \left(\int_0^\zeta \Delta(\lambda,y(\lambda))dw(\lambda)\right)d\zeta, \quad \eta\in[0b].$$

3. Existence and uniqueness results

Theorem 3.1. *Assume that Assumption 1 hold. Then the system* (1) *has a unique solution and following inequality is satisfied*

$$K := 2\left[E_{2q}\left(\|M\|^2(b+\kappa)^{2q}\right)\right]^2\left[\left(\frac{e^{-Nb}-1}{Nb}\right)^2 + 4b\left(\frac{e^{-2Nb}-1}{2Nb}\right)\right] < 1.$$

Proof. Let $x,y\in R^n$. From Assumption 1 for each $\eta\in[0,b]$, we have

$$|(\mathcal{P}x)(\eta)-(\mathcal{P}y)(\eta)|^2 \le 2\left|\int_0^\eta \cos_{\kappa,q} M(\eta-\kappa-\zeta)^q[F(\zeta,x(\zeta))-F(\zeta,y(\zeta))]d\zeta\right|^2$$
$$+2\left|\int_0^\eta \cos_{\kappa,q} M(\eta-\kappa-\zeta)^q\left(\int_0^\zeta[\Delta(\lambda,x(\lambda))-\Delta(\lambda,y(\lambda))]dw(\lambda)\right)d\zeta\right|^2.$$

Multiply by $e^{-2N\eta}$ on both sides, we get

$$e^{-2N\eta}|(\mathcal{P}x)(\eta)-(\mathcal{P}y)(\eta)|^2$$
$$\le 2\left|\int_0^\eta \cos_{\kappa,q} M(\eta-\kappa-\zeta)^q e^{-N\zeta}[F(\zeta,x(\zeta))-F(\zeta,y(\zeta))]d\zeta\right|^2$$
$$+2\left|\int_0^\eta \cos_{\kappa,q} M(\eta-\kappa-\zeta)^q e^{-N\zeta}\left(\int_0^\zeta[\Delta(\lambda,x(\lambda))-\Delta(\lambda,y(\lambda))]dw(\lambda)\right)d\zeta\right|^2 := 2[(i)+(ii)]. \tag{3}$$

First, we estimate (i):

$$\left|\int_0^\eta \cos_{\kappa,q} M(\eta-\kappa-\zeta)^q e^{-N\zeta}[F(\zeta,x(\zeta))-F(\zeta,y(\zeta))]d\zeta\right|^2$$
$$\le\left(\int_0^\eta|\cos_{\kappa,q} M(\eta-\kappa-\zeta)^q|^2 e^{-N(\eta-\zeta)}d\zeta\right)$$
$$\times\left(\int_0^\eta e^{-N(\eta-\zeta)}e^{-2N\zeta}|F(\zeta,x(\zeta))-F(\zeta,y(\zeta))|^2 d\zeta\right)$$
$$\le\left[E_{2q}\left(\|M\|^2(b+\kappa)^{2q}\right)\right]^2\left(\int_0^\eta e^{-N(\eta-\zeta)}d\zeta\right)$$
$$\times\left(\int_0^\eta e^{-N(\eta-\zeta)}d\zeta\right)e^{-2N\eta}|F(\eta,x(\eta))-F(\eta,y(\eta))|^2$$
$$\le\left[E_{2q}\left(\|M\|^2(b+\kappa)^{2q}\right)\right]^2\left(\int_0^\eta e^{-N(\eta-\zeta)}d\zeta\right)^2 E\|x(\eta)-y(\eta)\|^2$$
$$\le\left[E_{2q}\left(\|M\|^2(b+\kappa)^{2q}\right)\right]^2\left(\frac{e^{-Nb}-1}{Nb}\right)^2 E\|x(\eta)-y(\eta)\|^2.$$

By Burkholder-Davis-Gundy inequality and Assumption 1, the estimate for (ii) is given by

$$\left|\int_0^{\eta} \cos_{\kappa,q} M(\eta-\kappa-\zeta)^q e^{-N\zeta}\left(\int_0^{\zeta}[\Delta(\lambda,x(\lambda))-\Delta(\lambda,y(\lambda))]dw(\lambda)\right)d\zeta\right|^2$$

$$\le 4b\int_0^{\eta}|\cos_{\kappa,q} M(b-\kappa-\zeta)^q|^2 e^{-2N(b-\zeta)}e^{-2N\zeta}|\Delta(\zeta,x(\zeta))-\Delta(\zeta,y(\zeta))|^2 d\zeta$$

$$\le 4b\left[E_{2q}\left(\|M\|^2(b+\kappa)^{2q}\right)\right]^2\left(\int_0^{b} e^{-2N(b-\zeta)}d\zeta\right)e^{-2N\eta}|\Delta(\eta,x(\eta))-\Delta(\eta,y(\eta))|^2$$

$$\le 4b\left[E_{2q}\left(\|M\|^2(b+\kappa)^{2q}\right)\right]^2\left(\frac{e^{-2Nb}-1}{2Nb}\right)E\|x(\eta)-y(\eta)\|^2.$$

From the above two estimates of (i) and (ii), Eq. (3) becomes

$$E\|(\mathcal{P}x)(\eta)-(\mathcal{P}y)(\eta)\|^2$$

$$\le 2\left[E_{2q}\left(\|M\|^2(b+\kappa)^{2q}\right)\right]^2\left(\frac{e^{-Nb}-1}{Nb}\right)^2 E\|x(\eta)-y(\eta)\|^2$$

$$+8b\left[E_{2q}\left(\|M\|^2(b+\kappa)^{2q}\right)\right]^2\left(\frac{e^{-2Nb}-1}{2Nb}\right)E\|x(\eta)-y(\eta)\|^2$$

$$\le 2\left[E_{2q}\left(\|M\|^2(b+\kappa)^{2q}\right)\right]^2\left[\left(\frac{e^{-Nb}-1}{Nb}\right)^2+4b\left(\frac{e^{-2Nb}-1}{2Nb}\right)\right]E\|x(\eta)-y(\eta)\|^2.$$

This implies that

$$E\|(\mathcal{P}x)(\eta)-(\mathcal{P}y)(\eta)\|^2\le KE\|x(\eta)-y(\eta)\|^2.$$

Hence, from statement of the Theorem 3.1, the nonlinear operator $(\mathcal{P})$ is a contraction. Hence the nonlinear operator $(\mathcal{P})$ has a unique solution $y(\cdot)\in R^n$, which is nothing but solution of Eq. (1). Hence the proof. □

4. Finite-time stability

Theorem 4.1. *If Assumption 1 hold and provided that*

$$5\left[K_1+|M^{-1}|^2\left(K_3+K_2-2\sqrt{K_3}\sqrt{K_2}\right)+\kappa^2K_2\right]E_{q,1}\left[5\frac{b^{2-q}}{2-q}K_2\Gamma(q)\eta^q\right]\le\frac{\varepsilon}{\delta}.$$

Then the system (1) is finite-time stable in mean square.

Proof. By multiplying $e^{-2N\eta}$ on both sides of the solution of system (1), we derive

$$
\begin{aligned}
e^{-2N\eta}|y(\eta)|^2 \leq\; & 5\left|\left(\cos_{\kappa,q} M\eta^q\right)e^{-N\eta}\psi(-\kappa)\right|^2 + 5\left|M^{-1}\left(\sin_{\kappa,q} M(\eta-\kappa)^q\right)e^{-N\eta}\psi'(0)\right|^2 \\
& +5\left|\int_{-\kappa}^{0} \cos_{\kappa,q} M(\eta-\kappa-\zeta)^q e^{-N\zeta}\psi'(\zeta)d\zeta\right|^2 \\
& +5\left|\int_{0}^{\eta} \cos_{\kappa,q} M(\eta-\kappa-\zeta)^q e^{-N\zeta}F(\zeta,y(\zeta))d\zeta\right|^2 \\
& +5\left|\int_{0}^{\eta} \cos_{\kappa,q} M(\eta-\kappa-\zeta)^q e^{-N(\eta-\zeta)}e^{-N\zeta}\left(\int_0^{\zeta}\Delta(\lambda,y(\lambda))dw(\lambda)\right)d\zeta\right|^2 \\
\leq\; & 5\left|\left(\cos_{\kappa,q} M\eta^q\right)\right|^2 e^{-2N\eta}|\psi(-\kappa)|^2 \\
& +5\left|M^{-1}\right|^2\left|\left(\sin_{\kappa,q} M(\eta-\kappa)^q\right)\right|^2 e^{-2N\eta}|\psi'(0)|^2 \\
& +5\kappa^2\left|\cos_{\kappa,q} M(\eta+\kappa)^q\right|^2 e^{-2N\eta}|\psi'(\eta)|^2 + 5\left(\int_0^{\eta}(\eta-\zeta)^{q-1}d\zeta\right) \\
& \times\int_0^{\eta}(\eta-\zeta)^{1-q}\left|\cos_{\kappa,q} M(\eta-\kappa-\zeta)^q\right|^2 e^{-2N\zeta}|F(\zeta,y(\zeta)) \\
& -F(\zeta,0)|^2 d\zeta + 20\left(\int_0^{\eta}(\eta-\zeta)^{q-1}d\zeta\right) \\
& \times\int_0^{\eta}(\eta-\zeta)^{1-q}\left|\cos_{\kappa,q} M(\eta-\kappa-\zeta)^q\right|^2 e^{-2N\zeta}|\Delta(\zeta,y(\zeta)-\Delta(\zeta,0)|^2 d\zeta
\end{aligned}
$$

$$
\begin{aligned}
\|y(\eta)\|^2 \leq\; & 5\left[E_{2q}\left(\|M\|^2 b^{2q}\right)\right]^2\|\psi\|^2 + 5\left|M^{-1}\right|^2\left(\left[E_q(\|M\|(b+\kappa)^q)\right]^2\right. \\
& +\left[E_{2q}\left(\|M\|^2(b+\kappa)^{2q}\right)\right]^2 - 2E_q(\|M\|(b+\kappa)^q)E_{2q}\left(\|M\|^2(b+\kappa)^{2q}\right)\left.\right)\|\psi'\|^2 \\
& +5\kappa^2\left[E_{2q}\left(\|M\|^2(b+\kappa)^{2q}\right)\right]^2\|\psi'\|^2 \\
& +5\frac{b^{2-q}}{2-q}\left[E_{2q}\left(\|M\|^2(b+\kappa)^{2q}\right)\right]^2\left(\int_0^{\eta}(\eta-\zeta)^{q-1}\|y(\zeta)\|^2 d\zeta\right) \\
& +20\frac{b^{2-q}}{2-q}\left[E_{2q}\left(\|M\|^2(b+\kappa)^{2q}\right)\right]^2\left(\int_0^{\eta}(\eta-\zeta)^{q-1}\|y(\zeta)\|^2 d\zeta\right) \\
\leq\; & 5\left\{K_1\delta_1 + \left|M^{-1}\right|^2\left(K_3+K_2-2\sqrt{K_3}\sqrt{K_2}\right)\delta_1 + \kappa^2 K_2\delta_1\right. \\
& \left.+\frac{b^{2-q}}{2-q}K_2\left(\int_0^{\eta}(\eta-\zeta)^{q-1}\|y(\zeta)\|^2 d\zeta\right) + 4\frac{b^{2-q}}{2-q}K_2\left(\int_0^{\eta}(\eta-\zeta)^{q-1}\|y(\zeta)\|^2 d\zeta\right)\right\} \\
\leq\; & 5\left\{\left[K_1 + \left|M^{-1}\right|^2\left(K_3+K_2-2\sqrt{K_3}\sqrt{K_2}\right) + \kappa^2 K_2\right]\delta\right. \\
& +\frac{b^{2-q}}{2-q}K_2\left(\int_0^{\eta}(\eta-\zeta)^{q-1}\|y(\zeta)\|^2 d\zeta\right) \\
& \left.+4\frac{b^{2-q}}{2-q}K_2\left(\int_0^{\eta}(\eta-\zeta)^{q-1}\|y(\zeta)\|^2 d\zeta\right)\right\}.
\end{aligned}
$$

According to Lemma 2.1, let as take

$$b(\eta) = 5\left[K_1 + |M^{-1}|^2\left(K_3 + K_2 - 2\sqrt{K_3}\sqrt{K_2}\right) + \kappa^2 K_2\right]\delta$$

and

$$h(\eta) = 5\frac{b^{2-q}}{2-q}K_2.$$

Moreover, $b(\eta)$ is a nondecreasing function on $[0, b]$, then

$$v(\eta) \le b(\eta)E_{q,1}[h(\eta)\Gamma(q)\eta^q]$$

$$\|y(\eta)\|^2 \le 5\left[K_1 + |M^{-1}|^2\left(K_3 + K_2 - 2\sqrt{K_3}\sqrt{K_2}\right) + \kappa^2 K_2\right]\delta E_{q,1}\left[5\frac{b^{2-q}}{2-q}K_2\Gamma(q)\eta^q\right]$$

Then from the statement of Theorem 4.1, we get

$$\|y(\eta)\|^2 \le \varepsilon.$$

Hence the system (1) is finite-time stable in mean square. Hence the proof. □

Remark 4.1. *By using fixed-point rule, existence and uniqueness of solution, and controllability results have been investigated in* [27]. *Some well-known results on relative controllability of semilinear delay differential system with linear parts defined by permutable matrices are studied in* [28]. *In this chapter, we proved some new results of finite-time stability criteria in finite-dimensional space by employing Generalized Gronwall-Bellman inequality and suitable assumption on nonlinear terms.*

5. An example

Consider Eq. (1) in the below matrix form:

$$\begin{cases} {}^C D^{0.5}_{-0.75^+}\left({}^C D^{0.5}_{-0.75^+} y_1\right)(\eta) + 0.01 y_1(\eta - 0.75) = -(3-\eta)\dfrac{y_1^2(\eta)}{1-\eta} + \displaystyle\int_0^\eta (\zeta y_1(\zeta)\Delta_1 dB_1(\zeta), \\ y_1(\eta) = 2\eta, y_1'(\eta) = 2, \quad \eta \in [-0.75, 0]; \\ {}^C D^{0.5}_{-0.75^+}\left({}^C D^{0.5}_{-0.75^+} y_2\right)(\eta) + 0.01 y_2(\eta - 0.75) = -(3-\eta)\dfrac{y_2^2(\eta)}{1-\eta} + \displaystyle\int_0^\eta (\zeta y_2(\zeta)\Delta_2 dB_s(\zeta), \\ y_2(\eta) = 4\eta, y_2'(\eta) = 4, \quad \eta \in [-0.75, 0], \end{cases} \tag{4}$$

where $q = 0.5, \quad \kappa = 0.75, \Delta_1 = 0.3, \quad \Delta_2 = 0.5$

$$A = \begin{pmatrix} 0.1 & 0 \\ 0 & 0.1 \end{pmatrix}, \quad F(\eta, y(\eta)) = \begin{pmatrix} -(3-\eta)\dfrac{y_1^2(\eta)}{1-\eta}e^{2N\eta} \\ -(3-\eta)\dfrac{y_2^2(\eta)}{1-\eta}e^{2N\eta} \end{pmatrix}.$$

$$\Delta(\eta, y(\eta)) = \begin{pmatrix} -\eta y_1(\eta)e^{2N\eta}\sigma_1 dB_1 \\ -\eta y_2(\eta)e^{2N\eta}\sigma_2 dB_2 \end{pmatrix}, \quad \psi(\eta) = \begin{pmatrix} 2\eta \\ 4\eta \end{pmatrix}, \quad \psi'(\eta) = \begin{pmatrix} 2 \\ 4 \end{pmatrix}.$$

Further, we have the following fractional delayed cos matrices:

$$\cos_{0.75,0.65}\left(M\eta^{0.65}\right)=\begin{cases}\Theta, & -\infty<\eta<-0.75,\\ I, & -0.75\leq\eta<0,\\ I-M^2\dfrac{\eta^{1.3}}{\Gamma(2.3)}, & 0\leq\eta<0.75,\\ I-M^2\dfrac{\eta^{1.3}}{\Gamma(2.3)}+M^4\dfrac{(\eta-0.75)^{2.6}}{\Gamma(3.6)}, & 0.75\leq\eta<1.5.\end{cases} \tag{5}$$

From Eq.(5) and using basic calculation, one can get $\left[E_{2q}\left(\|M\|^2(b+\kappa)^{2q}\right)\right]^2=0.9712$, $\left(\frac{e^{-Nb}-1}{Nb}\right)^2=0.2683$ and $4b\left(\frac{e^{-2Nb}-1}{2Nb}\right)=-1.9004$. Using the above-obtained values, one can easily verify that

$$2\left[E_{2q}\left(\|M\|^2(b+\kappa)^{2q}\right)\right]^2\left[\left(\frac{e^{-Nb}-1}{Nb}\right)^2+4b\left(\frac{e^{-2Nb}-1}{2Nb}\right)\right]<1.$$

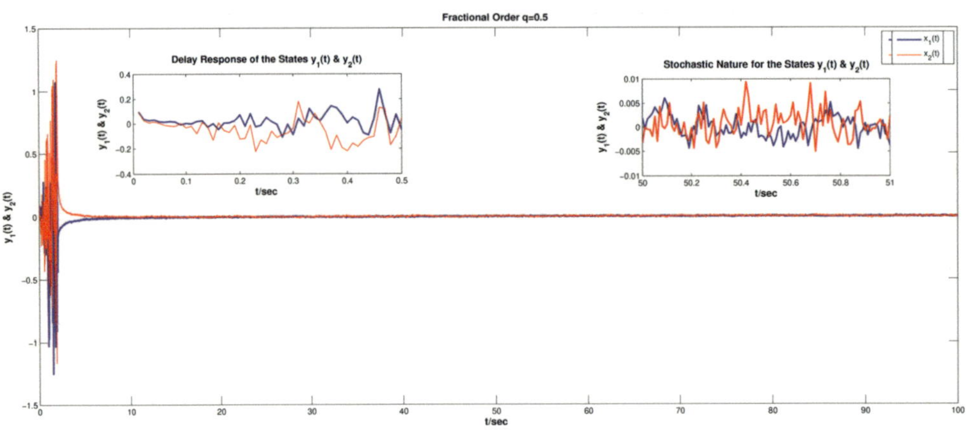

Figure 1.
The system (4) is stable at $q = 0.5$.

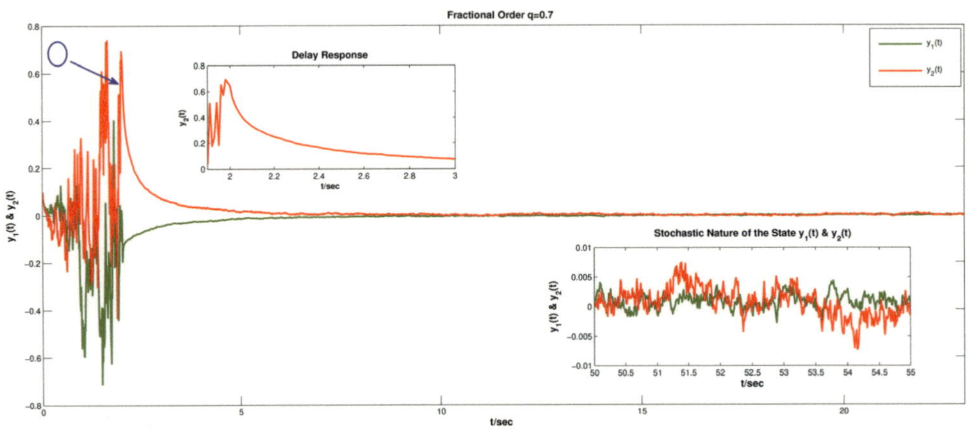

Figure 2.
The system (4) is stable at $q = 0.7$.

Hence we verified Theorem 3.1. Further, it is easy to verify that for any $x(\eta), y(\eta) \in R^2$.

$$e^{-2N\eta}|F(\eta, x(\eta)) - F(\eta, y(\eta))|^2 \leq -(3-\eta)\mathbb{E}\|x(\eta) - y(\eta)\|^2$$
$$e^{-2N\eta}|\Delta(\eta, x(\eta)) - \Delta(\eta, y(\eta))|^2 \leq -0.5\eta \ \mathbb{E}\|x(\eta) - y(\eta)\|^2$$

Hence, F and Δ satisfies Assumption 1. In **Figures 1** and **2**, we showed the stable response of the system (4) with fractional order $q = 0.5$ and $q = 0.7$, respectively. From the above verification, one can conclude that the system (4) is finite-time stable in mean square.

6. Conclusion

In this chapter, we have derived some meaningful and general results for finite-time stability of nonlinear fractional stochastic delay systems. Existence, uniqueness of solution and stability analysis of FSDS have been proved in finite-dimensional stochastic fractional higher-order differential system. Finally, a numerical simulation test is carried out to validate the obtained theoretical results. Derived result generalizes many existing results with integer and fractional-order systems.

AMS subject classifications (2010)

34A08; 43A15; 37C25; 37A50

Author details

Sathiyaraj Thambiayya[1], P. Balasubramaniam[2]*, K. Ratnavelu[3] and JinRong Wang[4]

1 Institute of Actuarial Science and Data Analytics, UCSI University, Kuala Lumpur, Malaysia

2 Department of Mathematics, The Gandhigram Rural Institute (Deemed to be University), Gandhigram, Tamil Nadu, India

3 Institute of Computer Science and Digital Innovation, UCSI University, Kuala Lumpur, Malaysia

4 Department of Mathematics, Guizhou University, Guiyang, China

*Address all correspondence to: balugru@gmail.com

References

[1] Hilfer R. Applications of Fractional Calculus in Physics. Singapore: World Scientific; 2000

[2] Oldham KB. Fractional differential equations in electrochemistry. Advances in Engineering Software. 2010;**41**: 1171-1183

[3] Magin RL. Fractional calculus models of complex dynamics in biological tissues. Computers and Mathematics with Applications. 2010;**59**:1586-1593

[4] Ortigueira MD. Fractional Calculus for Scientists and Engineers. New York: Springer Science & Business; 2011

[5] Kilbas AA, Srivastava HM, Trujillo JJ. Theory and Applications of Fractional Differential Equations. Amsterdam: North-Holland Mathematics Studies, Elsevier; 2006

[6] Nieto JJ. Solvability of an implicit fractional integral equation via a measure of noncompactness argument. Acta Mathematics Scientia. 2017;**37**: 195-204

[7] Singh J, Kumar D, Nieto JJ. Analysis of an el nino-southern oscillation model with a new fractional derivative. Chaos, Solitons and Fractals. 2017;**99**:109-115

[8] Tian Y, Nieto JJ. The applications of critical-point theory discontinuous fractional-order differential equations. Proceedings of the Edinburgh Mathematical Society. 2017;**60**: 1021-1051

[9] Mao X. Stochastic Differential Equations and Applications. Chichester: Horwood Publishing, Cambridge; 1997

[10] Wu D. On the solution process for a stochastic fractional partial differential equation driven by space-time white noise. Statistics and Probability Letters. 2011;**81**:1161-1172

[11] Seo D, Lee H. Stationary waiting times in m-node tandem queues with production blocking. IEEE Transactions on Automatic Control. 2011;**56**:958-961

[12] Taheri M, Navaie K, Bastani M. On the outage probability of SIR-based power-controlled DS-CDMA networks with spatial Poisson traffic. IEEE Transactions on Vehicular Technology. 2010;**59**:499-506

[13] Applebaum D. Levy Processes and Stochastic Calculus. Cambridge: Cambridge University Press; 2009

[14] Rockner M, Zhang T. Stochastic evolution equations of jump type: Existence, uniqueness and large deviation principle. Potential Analysis. 2007;**26**:255-279

[15] Ahmed E, El-Sayed AMA, El-Saka HAA. Equilibrium points, stability and numerical solutions of fractional-order predatorprey and rabies models. Journal of Mathematics Analysis and Applications. 2007;**325**:542-553

[16] Gao X, Yu J. Chaos in the fractional order periodically forced complex duffing oscillators. Chaos, Solitons and Fractals. 2005;**24**:1097-1104

[17] Odibat ZM. Analytic study on linear systems of fractional differential equations. Computers and Mathematics with Applications. 2010; **59**:1171-1183

[18] Wang J, Zhou Y, Fečkan M. Nonlinear impulsive problems for fractional differential equations and Ulam stability. Computers and Mathematics with Applications. 2012; **64**:3389-3405

[19] Burton TA, Zhang B. Fractional equations and generalizations of Schaefers and Krasnoselskii's fixed point theorems. Nonllinear Analysis: Theory,

Methods and Applications. 2012;**75**: 6485-6495

[20] Balasubramaniam P, Sathiyaraj T, Priya K. Exponential stability of nonlinear fractional stochastic system with Poisson jumps. Stochastics. 2021; **93**:945-957

[21] Fečkan M, Sathiyaraj T, Wang JR. Synchronization of Butterfly fractional order chaotic system. Mathematics. 2020;**8**:446

[22] Ren Y, Jia X, Sakthivel R. The p-th moment stability of solutions to impulsive stochastic differential equations driven by G-Brownian motion. Applicable Analysis. 2017;**96**:988-1003

[23] Shen G, Sakthivel R, Ren Y, Mengyu L. Controllability and stability of fractional stochastic functional systems driven by Rosenblatt process. Collectanea Mathematica. 2020;**71**:63-82

[24] Sathiyaraj T, Wang JR, Balasubramaniam P. Ulam's stability of Hilfer fractional stochastic differential systems. The European Physical Journal Plus. 2019;**134**:605

[25] Liang C, Wang J, O'Regan D. Representation of a solution for a fractional linear system with pure delay. Applied Mathematics Letters. 2018;**77**: 72-78

[26] Ye H, Gao J, Ding Y. A generalized Gronwall inequality and its application to a fractional differential equation. Journal of Mathematical Analysis and Applications. 2007;**328**:1075-1081

[27] Sathiyaraj T, Balasubramaniam P. Fractional order stochastic dynamical systems with distributed delayed control and Poisson jumps. The European Physical Journal Special Topics. 2016;**225**:83-96

[28] Wang J, Luo Z, Fečkan M. Relative controllability of semilinear delay differential systems with linear parts defined by permutable matrices. European Journal of Control. 2017;**30**: 39-46

Chapter 3

Control of Supply Chains

Kannan Nilakantan

Abstract

This chapter aims to apply control principles in the discrete-time control of Supply Chains. The primary objective of the control is to keep the inventory levels (state variables) steady at their predetermined values and reduce any deviations to zero in the shortest possible time. The disturbances are induced by demand deviations from the planned/anticipated levels. The replenishment flows are the control variables. Thus, the control action is very similar to a "Linear Regulator with zero set-point". A novel development in this chapter is the use of direct Operator methods to solve the system Difference Equations, thereby obviating the need for Z-Transforms, block diagrams and transfer functions of classical control theory. This chapter provides a novel application of control theory as well as an easier method of solution.

Keywords: Supply Chain Dynamic Modeling, Supply Chain Control, Feedback Control, Linear Regulator Problem, Direct Operator Methods

1. Introduction

In the field of business, one of the most important constituents of a business system, and one which can give a business a cutting-edge over the others, is the supply chain (SC) of the business, and its effective operation and management.

A fundamental strategy associated with supply chains is that of a 'Responsive Chain'. A responsive chain focuses on its ability to respond quickly to demand changes and meet them within the shortest possible time. A responsive chain, by its very definition, has *necessarily* to be able to respond swiftly to unanticipated changes in demand, and this is determined largely by the dynamics of the system. Under normal circumstances, good responsiveness can be achieved through the maintenance of *adequate pipeline inventories* throughout the system, and one of the factors impacting this decision in a major way, is again the dynamics of the system. Hence, given the impact of the dynamics of the SC system on its design and operation as mentioned above, it is imperative to include the tenets of dynamic analysis in its design and operation. Thus system-dynamic methods are of significant value and utility in both SC design and SC operation. The design aspects pertain to the setting of inventory levels and the operational aspects to choosing the appropriate system controls for good operational performance.

Thus, we are led naturally to the use of system-dynamic methods and concepts of control theory for effective control of a supply chain. In the following sections we explore the dynamic modeling and analysis concepts in effectively controlling a supply chain that would yield good performance characteristics.

2. Supply chain dynamics and notation

Supply chain dynamics essentially deals with the dynamic behavior and temporal variation of the inventories and flows in the system over time when subjected to demand disturbances.

We now look at a simple three stage serial supply chain, a schematic diagram of which is given below in **Figure 1**. Each stage represents an inventory storage facility in the chain. Stage 1 represents the raw material input storage to a manufacturing plant, and stage 2 the finished goods inventory storage at the manufacturing plant. Stage 3 represents the finished goods warehouse (W/H) at the downstream end of the chain which meets the demand for the finished product, and from which material is shipped out to customers. The inventory levels at each stage are the state variables of the system, and the material flows between the stages the control variables.

In constructing a dynamic model of a supply chain system, we adopt the standard schematic diagram (**Figure 1**), system notation, and model variables and equations below, as commonly found in control theory literature [1–9].

We use a discrete time representation of the SC system, as this is more in keeping with the prevailing practices in the SC industry, wherein the inventory levels (state variables) are recorded at the end of each period or day. In most cases the state variable records are updated at the end of each day. Hence, we take our period to be a single day. However, we need to also emphasize that this need not always be so and would be dependent entirely on the convenience of the SC practitioners. And hence the state variables are recorded at epochs corresponding to the end of each period. The flow variables however are aggregated and taken to occur within and up to the end of each period.

The inventory and flow variables are subscripted to indicate the stages in the chain. We now give the detailed representation below:

$y_i(k)$: inventory at stage i of the chain at time epoch k

with: i = 1, representing (the raw material) the upstream end of the production facility.

i = 2 representing (the finished goods) the downstream end of the production facility.

i = 3 representing (the finished goods) the warehouse.

$q_i'(k)$: material flow in period (k-1,k] into stage i of the chain

$r_3'(k)$: finished goods demand at warehouse in period (k-1,k]

The dynamic equations of the system are written using deviation variables, as under:

$$x_i(k) = y_i(k) - y_i^0(k) \text{ for i} = 1, 2, 3$$

$$q_i(k) = q_i'(k) - q_i^0(k) \text{ for i} = 1, 2, 3$$

$$r_3(k) = r_3'(k) - r_3^0(k)$$

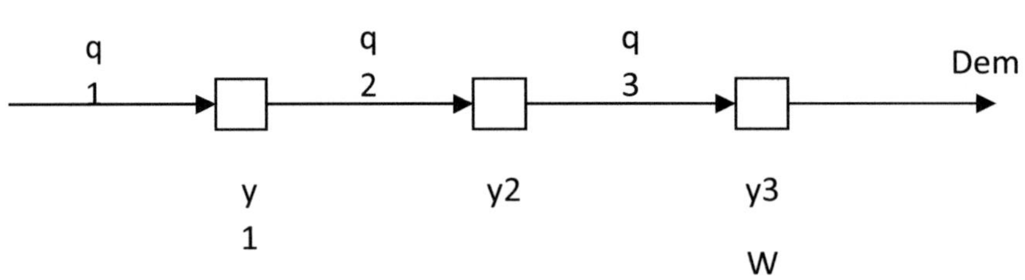

Figure 1.
A three-stage serial supply chain.

where,

$y_i^0(k)$ is the desired or planned inventory level at time k

$q_i^0(k)$ is the desired or planned flow in period (k-1,k]

$r_3^0(k)$ is the anticipated or forecasted demand at the warehouse in period (k-1,k], based on which the desired inventory levels in the chain have been set.

$x_i(k)$ is the inventory deviation at time k.

$q_i(k)$ is the material flow deviation in period (k-1,k].

$r_3(k)$ is the deviation in demand observed at the warehouse in (k-1, k].

Under consistent units (in equivalent units of finished goods) for all inventory and flows in the system, the dynamic equations of the system are written as:

$$x_i(k+1) = x_i(k) + q_i(k+1) - q_{i+1}(k+1) \quad (1)$$

where for i = 3, $q_4(k) = r_3(k)$, the demand outflow from the warehouse. The system behavior is controlled by the replenishment policies followed in the system.

Now, since the demand is a stochastic variable, it can be split into two components, viz., a mean demand component, and a stochastic component which represents the random variations over and above the mean demand. The mean demand is what is predicted using forecasting techniques and is the planned offtake and the demand that the system is designed to meet in the normal course. Since prediction can never be exact, the residual variation is the stochastic component.

In such stochastic systems, the demand has an additional stochastic term, which is a white noise term, given by $\varepsilon(k) \sim WN(0, \sigma^2)$ for all k. The sequence of stochastic disturbance components over each period is an independent and identically distributed sequence of $N(0, \sigma^2)$ random variables.

The standard initial conditions for the system are: $x_i(k) = 0$ for all $k \leq 0$ and $r_3(k) = 0$ for all $k \leq 1$, i.e. the system is at zero deviation at time k = 0, and the first deviation in demand is felt at the end of the first period, at k = 1.

The demand disturbance could be of the following Types:

1. a sudden shock demand increase, represented by a Dirac delta input function.

2. a sudden and sustained increase in demand, represented by a step input.

3. a demand with an increasing trend, represented by a ramp input function.

4. a demand with second order (quadratic), third order (cubic), or higher-order trends represented by higher-order polynomial input functions.

5. a demand with seasonality, represented by a sinusoidal input function.

6. A random component which is represented by a White Noise process.

The first five components pertain to and define *the mean demand disturbance*, while the last represents the *residual random variations over and above the mean demand disturbance*.

The demand can have any combination of, or even all of the above components in the mean demand, and, of course, the additional stochastic component represented by a White Noise process. Such demands are often seen in supply chain warehouses.

Thus, a demand disturbance at the downstream end (at the warehouse) provides the perturbation to the system. The system is controlled through regulation/control of the replenishment flows which are the control variables in the system.

Now from the above, we can see that the system response will have two components, the *Mean Response*, and a stochastic component of the response. It is the *mean response that characterizes the system behavior*, while the stochastic component is characterized by the inventory variance.

3. Dynamic supply chain performance metrics and control action triggers

3.1 Dynamic performance metrics

The performance metrics of a supply chain that would be of interest to us from a dynamic performance point of view and control system design are explained below.

Since the demand input to the system suddenly increases, we can expect the system response to lag the demand and to fluctuate, as the system scrambles to catch up with the increased demand. Accordingly, the key performance indicators that would be of importance and interest to us from the point of view of both design and operation are the following [1–9]:

a. Permanent depletion of the inventory level (the offset), if any.

b. The trough value or the lowest dip in the inventory levels (the undershoot).

c. The amplitude of fluctuations of the inventory.

d. The center-line about which fluctuations occur.

e. The fraction of time the inventory level stays depleted (in the negative region).

f. The limiting inventory variance.

The first indicator shows whether the system is able to catch up with the demand and whether it is ultimately restored to its original level, or not. The second indicator, the trough value, represents the lowest point or value that the inventory level is likely to touch, and impacts the *base stock levels* that would have to be carried to maintain uninterrupted material flows in the system, (adequate pipeline inventories). The third defines the magnitude of fluctuations of the inventory levels, and high values would naturally be undesirable; and we would like to damp them down to zero as quickly as possible. The fourth indicates the center-line about which fluctuations occur and is indicative of the *average inventory level*s, which we would like to keep at positive levels for comfortable operation. While the fifth can be taken to be a surrogate measure of the *stock-out risk*, and the higher the fraction of time in the negative region (with a depleted inventory level), the higher could be taken to be the *implied stock-out risk*. The last, the inventory variance is the variation that we could expect even after the system has been restored to its original operating levels and is commonly taken as a measure of the robustness of the system to random demand variations.

3.2 Control action triggers

The two most common triggers for initiation of replenishment control action in the system are [1–9]:

1. The inventory levels at the W/H, as found in logistic systems and warehouses,

2. The demand variations at the W/H, as in electronic-data-interchange systems.

Thus, the control flows into the W/H are set to a function of the latest available inventory deviations and latest available Demand deviations. Thus, we have,

$$q_3(k+1) = f[x_3(k-1), x_3(k-2), .., x_3(k-r); r_3(k-1), r_3(k-2), \ldots r_3(k-p))] \tag{2}$$

We discuss these points in more detail for a single stage system first.

3.3 Single stage control notation

To differentiate between the standard abbreviations in conventional control theory, we adopt the following notation used to indicate the type of control used.

P, PI, PID would represent Proportional, Proportional-integral, and Proportional-integral-derivative control as usual. Additionally, MA denotes a 'Moving Average' type of control (explained in detail subsequently).

An 'I' within parenthesis would represent 'inventory-triggered' control, while 'D' within parenthesis would denote a 'Demand-triggered' control. Also, 'ID' within parenthesis would denote a control which would have both inventory-triggered and Demand-triggered components. We give examples below.

Thus P(D) (pronounced as "P of D") denotes Demand-triggered Proportional control, PI(I) denotes Inventory-triggered PI control. Similarly, PID(ID) (pronounced as "PID of ID") denotes PID control with inventory-triggered and Demand-triggered components.

We also could have cases of multiple inventory triggers and multiple-demand triggers. These will be denoted as follows:

$X(I_n D_m)$ control would indicate a control of type X (P, PI, PID, MA, Composite) with n-inventory trigger terms and m-demand trigger terms. Thus, the control $PD(I_5 D_3)$ (pronounced as "PD of I5D3") would be Proportional-derivative control with 5 inventory triggers and 3 demand triggers.

We now first look at a single stage system below which we take to be the warehouse end of a SC.

4. Single stage control

4.1 The two response components

Since the demand has both a deterministic component as well as a stochastic component, the response of the system can also be broken down into two components, viz., a deterministic part which is *the mean response*, and a random or stochastic part characterized by the *inventory variance*. It is the mean response that portrays the behavior of the system and yields the performance indicators of the system. Whereas, the stochastic part, represents the random fluctuations that have to be accounted for even after the system is brought under full control.

From the above discussion we can see the close parallel with conventional feedback control theory. Here the information about the warehouse inventory (the state variable) is fed back to the system for initiating replenishment flow control action. Additionally, we can also have controls wherein the disturbance is also directly fed back to the system for initiation of control action.

4.2 Single stage system controls and transportation lags

In close parallel with classical feedback control theory, the control flows are functions of the state variables and the demand or input perturbation to the system. The types of functions used also closely parallel classical control theory. And hence we have P, PI, and PID controls.

Additionally, we also have Moving Average (MA) controls, wherein the control flow is set to a weighted Moving Average of the latest available inventory deviations of up to 'r' periods back. The parameter 'r' is termed the 'Order' of the Moving Average.

We first look at Proportional Controls.

Now, for Proportional Controls of the P(I) type, the control flows into the warehouse in stage 3 of the chain would be given by:

$$q_3(k+1) = K_3 x_3(k-1-l_3) \tag{3}$$

where l_3 is the transportation lag in the flows into the warehouse from the upstream unit of the system, i.e., the finished goods inventory at the manufacturing plant in our case. And K_3 is the constant of proportionality between the control flow and the *latest available inventory deviation* based on which the order is initiated (and hence 'Proportional' to the error in conventional control theory).

The development herein corresponds to the 'Regulator Problem' with 'set point' of zero. And hence control of a Responsive Chain parallels the regulator problem in conventional control theory.

In conventional modeling of SCs, the lag is taken as the number of periods strictly between the period of order initiation and the period of arrival of the consignment, not including the period of arrival of the consignment. Thus, the lag is taken to be zero if the replenishment consignment arrives in the period immediately succeeding the period of order initiation. Instantaneous replenishment is not envisaged and is very rare in SC contexts. Arrival of the consignment within the same period of order initiation is also not envisaged and is very rare in such contexts. The earliest arrival of an ordered consignment is taken to be the immediately succeeding period, for which we take the lag as zero.

This convention is based on what is normally followed in the industry and practice, as well as the literature on dynamic modeling of SCs.

Hence in our further development of dynamic models of SCs we take the lag as zero if an order initiated in period k i.e., in the interval (k – 1, k] arrives in period (k + 1), i.e., in the interval (k, k + 1].

For a transportation lag of 'l' periods, an order placed in the interval (k – 1, k] would arrive in the interval (k + l, k + l + 1].

Hence for the warehouse under *zero lag* the control flows for P(I) control would be set as:

$$q_3(k+1) = K_3 x_3(k-1) \tag{4}$$

Now it is to be noted that the order is initiated in period k i.e., in the interval (k – 1, k] based on the *latest fully observed inventory deviation* which in this case would be that at the start of period k, which is the inventory recorded at the end of period (k-1) i.e., $x_3(k-1)$. We take it that since the inflows in period k would be the aggregate of flows in the interval (k – 1, k], and the order is initiated *within* the period k (*and not at time point k*), the latest available fully observed inventory level would be $x_3(k-1)$ and not $x_3(k)$ because $x_3(k)$ would not be available to the order initiator within period k. The value of $x_3(k)$ is recorded at the closure of period k, after the cessation of all activities including the action of ordering, of period k. Thus, replenishment orders need to be initiated within and before the closure of a period and not at the end of a period.

Thus, warehouse records would be updated at the closure of the day's operations and would show the *closing inventory* and the replenishment orders placed *during* the day. The replenishment orders would have been placed based on the previous day's closing stock and would arrive during the course of the next day if the lag is zero.

This is the convention that we will follow in the further development of the models.

We next look at the next type of control, which is the PI(I) control.

For the Proportional-Integral (PI(I) type) control case with *zero lag*, the control flows at the warehouse would be set as under:

$$q_3(k+1) = K_3 x_3(k-1) + K_c \sum_{m=0}^{k-1} x_3(m) \tag{5}$$

where the second term is the integral term in our discrete-time system, and K_c is the proportionality constant (gain term) factor of the integral of the error.

Next we have the PID(I) control wherein the control flows into the warehouse under *zero lag* would be set to:

$$q_3(k+1) = K_3 x_3(k-1) + K_c \sum_{m=0}^{k-1} x_3(m) + K_d(x_3(k-1) - x_3(k-2)) \tag{6}$$

where the last term represents the derivative term in our discrete-time system, and K_d is the proportionality constant factor (gain term) of the derivative component of the control.

Next we have the Moving Average (MA) type of control (MA(I) type), for which the control flows into the warehouse under *zero lag* would be set to:

$$q_3(k+1) = K_3^1 x_3(k-1) + K_3^2 x(k-2) + K_3^3 x(k-3) + \dots \dots + K_3^r x(k-r) \tag{7}$$

where the r is the order of the moving average, the Ks are the control parameters (the weights) of the MA terms. Thus, the control flow is set to a weighted moving average of the latest available fully observed inventory levels up to r period back.

The above controls discussed above are the conventional inventory-triggered schemes. In all these types of controls, we could additionally have demand-triggered terms also, like the PI(ID), PID(ID), MA(ID) etc.

We discuss some of them below for single stage systems.

5. Solution of single stage systems

Firstly, we note herein that in solving for the response of a supply chain system, we will not use the Z-transform nor block diagrams and transfer functions as in conventional control theory. Rather we will work directly on the system difference equation in the time domain itself. And instead, we will use direct Operator methods to obtain the system solution and response (rather than the transformed equations and inverse transforms). This is one of the advantages of this modeling paradigm.

Another advantage of the type of discrete time modeling taken up here is that *transportation lags do not result in differential-delay equations* as in conventional continuous-time control theory. Rather, transportation lags would increase the order of the resulting system difference equation, which is expected to be easier to solve than differential-delay equations.

We now take up the simplest form of control which is the P(I) control and illustrate the formulation of the system equation and its solution method.

5.1 P(I) control under zero lag

The control is an inventory-triggered Proportional control. And we take the Replenishment Lag = 0 in the simplest case.

The flow balance equation for the warehouse is given by:

$$x_3(k+1) = x_3(k) + q_3(k+1) - r_3(k+1) \tag{8}$$

The control flows into the warehouse are given by:

$$q_3(k+1) = K_3 x_3(k-1) \tag{9}$$

It is to be noted that the value of $K_3 < 0$, since the control flows and inventory deviation have to be of opposite sign, i.e., when the inventory deviation is (−)ve (inventory is at a depleted level) then the flow deviation has to be (+)ve to enable more/extra flow into the warehouse to make up the inventory shortfall. Likewise, if the inventory deviation is (+)ve (extra inventory in stock) then the flow deviation has to be (−)ve (i.e., reduced flow) to reduce the inflow into the warehouse to maintain inventory levels.

Thus, we can clearly see that the controls are of the 'feedback' type, and they seek to keep the inventory deviation at zero level, which is just the' Linear Regulator with zero set-point' in standard control theory.

Thus, substituting for the control flow into the flow balance eqn. Above, yields the system-dynamic eqn. For the warehouse as:

$$x_3(k+1) \equiv x_3(k) + K_3 x_3(k-1) - r_3(k+1) \tag{10}$$

$$\text{Or equivalently}, x_3(k+1) - x_3(k) - K_3 x_3(k-1) \equiv -r_3(k+1) \tag{11}$$

The above is the deterministic part of the system equation. Since demand is a stochastic variable with a stochastic component, the complete system equation is given by:

$$x_3(k+1) - x_3(k) - K_3 x(k-1) \equiv -r_3(k+1) - \varepsilon(k+1) \tag{12}$$

which has both parts. To solve the system equation completely, we split it into its two components and solve for each of the components separately. We hence solve the following two equations, one each for the deterministic part and the stochastic part.

$$x_3^{\det}(k+1) - x_3^{\det}(k) - K_3 x_3^{\det}(k-1) \equiv -r_3(k+1) \text{ for the deterministic part, and,} \tag{13}$$

$$x_3^{stoc}(k+1) - x_3^{stoc}(k) - K_3 x_3^{stoc}(k-1) \equiv -\varepsilon(k+1) \text{ for the stochastic part.} \tag{14}$$

The first is a deterministic Linear Difference Equation, and the second, a Stochastic Linear Difference Equation (SDE).

Both are second-order Linear Difference Equations (LDEs) in the state variable $x_3(k)$ and can be solved by direct Operator methods for any type of input or disturbance term on the RHS of the system LDE.

An excellent treatment of difference calculus and solution methods for LDEs is given in [10]. We follow the methods given therein.

In order to solve the LDE, we first introduce the Forward Shift Operator E as under: $Ex_3(k) \equiv x_3(k+1)$, with the important property that: $E[Ex_3(k)] = E^2x_3(k) = x_3(k+2)$. Before using the Operator, we first write the LDE in standard form as under: (the lowest time value is taken as k):

$$x_3(k+2) - x_3(k+1) - K_3x_3(k) \equiv -r_3(k+2) \tag{15}$$

Now using the forward Shift Operator E, we can write the LDE in Operator form as:

$$[E^2 - E - K_3\}x_3^{\text{det}}(k) \equiv -r_3(k+2) \text{ for the deterministic part, and,} \tag{16}$$

$$[E^2 - E - K_3\}x_3^{stoc}(k) \equiv -\varepsilon(k+2) \text{ for the stochastic part.} \tag{17}$$

5.1.1 The mean response: solution of the deterministic LDE

We first look at the deterministic part of the solution below, which will yield the *mean response*. For notational convenience, we drop the superscript on $x_3(k)$.

$$[\mathrm{E}^2 - \mathrm{E} - \mathrm{K}_3]\mathrm{x}_3(\mathrm{k}) \equiv -\mathrm{r}_3(\mathrm{k}+2)$$

Now this is an LDE of order two (the order being the highest power of the Operator E).

We write the Characteristic Equation of the LDE as [10]:

$$\alpha^2 - \alpha - K_3 = 0 \tag{18}$$

to determine the characteristic roots of the LHS Operator.

$$\text{From which we get the roots as : } \alpha_{1/2} = \frac{1 \pm \sqrt{1+4K_3}}{2}. \tag{19}$$

The stability of the system is entirely controlled by the roots of the LHS Operator. And elementary analysis leads to the following stability conditions:

$K_3 \geq 0$: instability

$-1/4 \leq K_3 < 0$: stability with non-oscillatory response

$-1 \leq K_3 < -1/4$: stability with oscillatory behavior

$K_3 < -1$: instability with oscillatory behavior

Now we take up the solution of the system LDE for a unit step increase in demand, i.e.,

$$r_3(k+1) = 1, \forall k \geq 0 \tag{20}$$

Substituting for the demand disturbance in the system equation yields the LDE:

$$[E^2 - E - K_3]x_3(k) \equiv -1 \tag{21}$$

which we can call the "Original Non-Homogeneous Eqn." (O-NHE).

We first look at the solution of the homogeneous LDE (i.e., with RHS = 0). The homogeneous LDE is:

$$\left[E^2 - E - K_3\right]x_3(k) \equiv 0 \tag{22}$$

which upon factoring the LHS Operator can be written as: (λ_1, λ_2 being the distinct roots of the LHS Operator):

$$[(E - \lambda_1)(E - \lambda_2)]x_3(k) \equiv 0 \tag{23}$$

which has the solution as: $x_3(k) \equiv C_1\lambda_1^k + C_2\lambda_2^k$ for the case of distinct roots.

For a repeated root, the solution is given by: $x_3(k) \equiv (C_0 + C_1k)\lambda^k$ for a repeated root of algebraic multiplicity two.

Now that we have the solution of the homogeneous LDE, we next look for a particular solution of the Original Non-Homogeneous LDE (O-NHE).

A standard method of solution of the Non-homogeneous eqn. is by the 'Annihilator Method' [10].

We look for the Operator that annihilates the RHS terms of the O-NHE, say A (E). Then operating by the Annihilator on both sides of the O-NHE yields:

$$A(E)[(E - \lambda_1)(E - \lambda_2)]x_3(k) \equiv A(E)r_3(k+2) \equiv 0 \tag{24}$$

which is a homogeneous LDE albeit of a higher order, but which can be solved by factorizing the LHS Operator. As an example, in our case of a unit step disturbance, $r_3(k) \equiv -1, \forall k \geq 1$.

And the Annihilator is given by:

$$A(E) \equiv (E - 1), \because (E - 1)r_3(k) \equiv -(E - 1)(1) = 0$$

And hence the equivalent Homogeneous LDE is given by:

$$[(E - 1)(E - \lambda_1)(E - \lambda_2)]x_3(k) \equiv 0$$

which has the solution:

$$x_3(k) \equiv D\left(1^k\right) + C_1\lambda_1^k + C_2\lambda_2^k \equiv D + C_1\lambda_1^k + C_2\lambda_2^k \text{ for the case of distinct roots.} \tag{25}$$

$$x_3(k) \equiv D(1)^k + (C_0 + C_1k)\lambda^k \text{ for the case of repeated roots} \tag{26}$$

Now we note that the O-NHE being of order two, will admit only two undetermined constants. The above solution, however, has three undetermined constants. The third was introduced by us due to the Annihilator. Hence to determine the extra constant D in the solution, we substitute the solution into the O-NHE to determine D. Thus, and noting that the terms involving the roots of the LHS operator of the O-NHE are precisely the homogeneous solution terms of the O-NHE, we only need substitute the extra terms introduced by the Annihilator into the O-NHE. Hence, we have:

$$[(E - \lambda_1)(E - \lambda_2)]D \equiv \left[E^2 - E - K_3\right]D = -1$$

Noting that E(D) = D itself ($\because E(D) \equiv D(k+1) \equiv D, \forall k$), we obtain $D(k+2) - D(k+1) - K_3D \equiv D - D - K_3D = -1$ from which we readily obtain

$$D = 1/K_3 \text{ which is } (-)\text{ve due to the } (-)\text{ve sign of } K_3. \tag{27}$$

Thus, the full solution is given by:

$$x_3(k) \equiv 1/K_3 + C_1\lambda_1^k + C_2\lambda_2^k \text{ for the case of distinct roots, and} \tag{28}$$

$$x_3(k) \equiv 1/K_3 + (C_0 + C_1 k)\lambda^k \text{ for the case of repeated roots.} \tag{29}$$

Now, for stable solutions, we will have $|\lambda_1| < 1$, $|\lambda_2| < 1$, and hence the terms involving the roots of the LHS Operator decay to zero for large k, leaving the 'Offset' term as $1/K_3$ which is the Offset value. The Offset value is $(-)$ve because of the $(-)$ve sign of K_3.

The Damping rate, which is the rate at which the fluctuations decay to zero are given by the magnitudes of the roots of the LHS Operator $\left(|\lambda_1|^k, |\lambda_2|^k\right)$, which can clearly be seen to decrease in magnitude with decrease in magnitude of $|K_3|$. However, the Offset value increases in magnitude with decrease of K_3.

We take the case of distinct roots first, say for a value of $K_3 = -1/2$, in the stable region.

Hence, we have the system equation as:

$$\left[E^2 - E - K_3\right]x_3(k) \equiv [E - E - (-1/2)]x_3(k) \equiv -1 \tag{30}$$

The roots of the LHS Operator are readily obtained as: $\lambda = (1/2) \pm (1/2)j, |\lambda_1| < 1, |\lambda_2| < 1, j = \sqrt{-1}$, which can be written in the form.

$\lambda = \rho e^{\pm j\theta} \equiv (1/\sqrt{2})e^{\pm j\pi/4}$, which hence yields the homogeneous solution as: $\{ACos(k\pi/4) + BSin(k\pi/4)\}(1/\sqrt{2})^k$, and hence the full solution as:

$$x_3(k) \equiv (1/K_3) + \{ACos(k\pi/4) + BSin(k\pi/4)\}\left(1/\sqrt{2}\right)^k \tag{31}$$

$$\equiv -2 + \{ACos(k\pi/4) + BSin(k\pi/4)\}\left(1/\sqrt{2}\right)^k, \tag{32}$$

which shows a Damping Rate of the Order of $O(1/\sqrt{2})^k = O(0.707)^k$, with an Offset of - 2 units. The response if plotted in **Figure 2**, and the 'Undershoot' is obtained as -2.5 units.

We next take a value of $K_3 = -1/4$ for the case of repeated roots, again in the stable region. The roots of the LHS Operator are obtained as: $\lambda_1 = \lambda_2 = \lambda = 1/2$, and the full solution as:

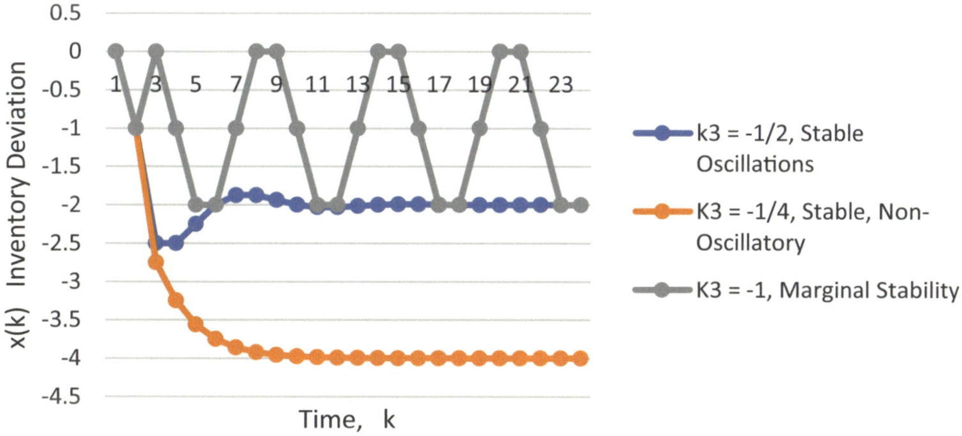

Figure 2.
P(I) control.

$$x_3(k) \equiv (1/K_3) + \{C_0 + C_1 k\}(1/2)^k \text{ which for } K_3 = -1/4 \text{ yields} \tag{33}$$

$$x_3(k) \equiv -4 + \{C_0 + C_1 k\}(1/2)^k, \tag{34}$$

which shows a Damping Rate of the Order of $O(1/2)^k = O(0.5)^k$, but with a high Offset value of - 4 units, and an Undershoot of - 4 units.

Thus, while we would like the oscillations to be damped out rapidly, this would compromise on the Offset value, which would impact the base stock requirements of the system.

Thus, for the practitioner, the implications are quite clear: *there is a trade-off between stability and rapid damping on the one hand, and the base stock requirements of the system (for low stock-out risk) on the other. The higher the damping (stability) required, the higher would be the base-stock requirements to keep stock-out risk low; and alternatively, the higher the (stability) damping achieved, the higher would be the stock-out risk at fixed base-stock levels.*

We next obtain the undetermined constants in the general solutions above, using the Initial Conditions (ICs) of the system, as under:

The standard ICs of the system are as: $\{x_3(k) \equiv 0, r_3(k) \equiv 0, \forall k \leq 0\}$.

Now the system LDE: $[E^2 - E - K_3]x_3(k) \equiv -r(k+2) \equiv -1, \forall k \geq 0$ is valid for $\forall k \geq 0$.

And hence the ICs for our warehouse system can be obtained from the system equation itself using the standard system ICs and yields: $\{x_3(0) = 0, x_3(1) = -1\}$. Substituting these ICs into the solutions above yields the full solutions as:

$$\text{For } K_3 = -1/2 : x_3(k) \equiv -2 + 2\text{Cos}(k\pi/4)\left(1/\sqrt{2}\right)^k, k \geq 0 \tag{35}$$

$$\text{For } K_3 = -1/4 : x_3(k) \equiv -4 + (4 + 2k)(1/2)^k, k \geq 0. \tag{36}$$

We additionally examine the case for $K_3 = -1$ (the maximum possible magnitude for stability), the response is given by

$x_3(k) \equiv -1 + [A\text{Cos}(k\pi/3) + B\text{Sin}(k\pi/3)](1)^k, k \geq 0$, which using the LDE ICs yields:

$$x_3(k) \equiv -1 + \text{Cos}(k\pi/3) + \left(1/\sqrt{3}\right)\text{Sin}(k\pi/3), k \geq 0, \tag{37}$$

which can be simplified to.

$$x_3(k) = \left\{-1 - \left(2/\sqrt{3}\right)\text{Sin}((k-1)\pi/3)\right\}H(k-1), k \geq 1, x_3(0) = 0. \tag{38}$$

where H(.) is the unit Heaviside step function and yields a sinusoidal pattern with a center-line of – 1 and constant amplitude of $2/\sqrt{3}$ and the maximum negative deviation in inventory, the undershoot, equal to -2.

This last case is that of *marginal stability characterized by constant amplitude perpetual oscillations which never die down to zero.*

The responses for the three cases above are plotted in **Figure 2**.

5.1.2 The limiting inventory variance: solution of the SDE

We next look at the determination of the limiting inventory variance, which is a measure of the variation that we could expect even after the system (mean inventory level) has been restored to its original value.

Also, for the behavior of the mean response as well as our inferences from it to be meaningful, it is necessary that the limiting inventory variance be bounded and finite. We can then expect the inventory levels to be within the band given by: $x_3(k)^{\text{det}} \pm 3\sqrt{\lim_{t\to\infty}\text{var}(x_3(k))}$, where $x_3(k)^{\text{det}}$ is the mean response that has been determined above from the deterministic system LDE.

In order to determine the limiting inventory variance, we make use of the stochastic component of the system equation and determine the stochastic part of the response. For our system the Stochastic LDE (or SDE) is as under:

$$[E^2 - E - K_3]x_3(k) \equiv -\varepsilon(k+2), \tag{39}$$

where the term on the RHS of the SDE is the random variation represented by a White Noise Process, with $\varepsilon(k) \sim WN(0, \sigma^2)$.

We can note that the LHS Operator is again the same as in the deterministic part of the system LDE that we have solved for above. Hence the roots of the LHS Operator remain unaltered in the SDE also.

Now following the method used in [11], we can note that if *unity is not a root of the LHS Operator*, then the SDE admits as a solution, an infinite weighted Moving Average representation in terms of the White Noise disturbance terms as under:

$$x_3(k)^{stoc} \equiv \sum_{l=0}^{k-1} \beta_l \varepsilon(k-l) \tag{40}$$

where the β_l s are the weights.

And hence the stochastic part of the solution of the system eqn. can be written as an infinite linear combination of the white noise disturbance terms.

Now since the individual white noise terms are Uncorrelated and Normal, i.e., with $\varepsilon(k) \sim N(0, \sigma^2)$, and, with $Cov(\varepsilon(i), \varepsilon(j)) = \delta_{ij}\sigma^2, \delta_{ij} = \begin{bmatrix} 0, i \neq j \\ 1, i = j \end{bmatrix}$, the limiting variance of x(k) is given by:

$$\lim_{k\to\infty}\text{var}(x_3(k)) = \lim_{k\to\infty}\text{var}\left(\left\{\sum_{l=0}^{k-1}\beta_l\varepsilon(k-l)\right\}\right) = \lim_{k\to\infty}\left\{\sum_{l=0}^{k}\beta_l^2\right\}\sigma^2 \tag{41}$$

In order to solve for the weighting terms, the βs, we substitute the solution into the system SDE above, as under:

$$[E^2 - E - K_3]x_3(k)^{stoc} \equiv -\varepsilon(k+2), \text{valid for all } k \geq 0, \tag{42}$$

which is

$$[E^2 - E - K_3]\left\{\sum_{l=0}^{k-1}\beta_l\varepsilon(k-l)\right\} \equiv -\varepsilon(k+2), k \geq 0. \tag{43}$$

Another and more convenient way to write the SDE is to use the Backward Shift Operator L, defined by:

$$L\text{x(k)} = \text{x(k}-1), \text{i.e.,} L = E^{-1}, E = L^{-1}. \tag{44}$$

The SDE for the βs becomes:

$$[1 - L - K_3L^2]\left\{\sum_{l=0}^{k-1}\beta_l\varepsilon(k-l)\right\} \equiv -\varepsilon(k), k \geq 0, \tag{45}$$

which is

$$\left[1-L-K_3L^2\right]\{\beta_0\varepsilon(k)+\beta_1\varepsilon(k-1)+\beta_2\varepsilon(k-2)+\beta_3\varepsilon(k-3)\ldots\}\equiv-\varepsilon(k),k\geq 0 \tag{46}$$

Now comparing coefficients of ε(k) for each k, yields the system of equations as below:

LDE term	Coefft of ε(k)	Coefft of ε(k-1)	Coefft of ε(k-2)	Coefft of ε(k-3)	Coefft of ε(k-4)	Coefft of ε(k-5)	….	Coefft of ε(k)
1	β_0	β_1	β_2	β_3	β_4	β_5	…..	β_k
- L	0	$-\beta_0$	$-\beta_1$	$-\beta_2$	$-\beta_3$	$-\beta_4$	…..	$-\beta_{k-1}$
$-K_3L^2$	0	0	$-K_3\beta_0$	$-K_3\beta_1$	$-K_3\beta_2$	$-K_3\beta_3$	…..	$-K_3\beta_{k-2}$
RHS	-1	0	0	0	0	0	0	0

The above set of equations yields: $\beta_0=-1,\beta_1-\beta_0=0,\&,\beta_k-\beta_{k-1}-K_3\beta_{k-2}=0,\forall k\geq 2$, which is an LDE for the βs, as.

$$\left[E^2-E-K_3\right]\beta_k\equiv 0,\forall k\geq 0,\text{with ICs}:\{\beta_0=-1=\beta_1\}. \tag{47}$$

We can note that the LHS Operator is the same as for the system LDE, and hence has the same characteristics and form of solution.

We first illustrate the computation for the stable case $K_3=-1/2$, for which the solution form has already been obtained as: $(A\text{Cos}(k\pi/4)+B\text{Sin}(k\pi/4))(1/\sqrt{2})^k$, and hence plugging in the ICs, we have the solution for βs as:

$$\beta_k\equiv(-\text{Cos}(k\pi/4)-\text{Sin}(k\pi/4))\left(1/\sqrt{2}\right)^k,k\geq 0.$$

To obtain the limiting inventory variance, we firstly note that:

$$\beta_k^2=(-\text{Cos}(k\pi/4)-\text{Sin}(k\pi/4))^2(1/2)^k=(1-2\text{Sin}k\pi/2)(1/2)^k\leq 3(1/2)^k. \tag{48}$$

And hence,

$$Lim_{k\to\infty}\sum_{l=0}^{k}\beta_l^2\leq 1+1+3(1/2)^2\left\{1+(1/2)+(1/2)^2+(1/2)^3+\ldots\ldots\right\}=2+3/2$$
$$=3.5. \tag{49}$$

Hence, we have: $Lim_{k\to\infty}\text{var}(x_3(k))\leq 3.5\sigma^2$, showing that the inventory variance is bounded and finite for this case.

We next illustrate the computation for the marginally stable case $K_3=-1$, for which the solution form has already been obtained as: $A\text{Cos}(k\pi/3)+B\text{Sin}(k\pi/3)$, and hence plugging in the ICs, we have the solution for the βs as:

$$\beta_k\equiv-\text{Cos}(k\pi/3)-\left(1/\sqrt{3}\right)\text{Sin}(k\pi/3)\equiv\left(2/\sqrt{3}\right)\text{Cos}(k\pi/3-\pi/6),k\geq 0 \tag{50}$$

And we can see from the above that the βs oscillate in value, from −1 to 0 to +1 infinitely often, i.e., the sequence {… 0, −1, −1, 0, +1,+1, 0 … ..} repeats infinitely often. And hence the series $\sum_{k=0}^{\infty}\beta_k^2$ diverges to infinity.

Hence in this case the limiting inventory variance is *not bounded* and *not finite*.

It can similarly be shown that the limiting inventory variance is not bounded for unstable solutions also.

Thus, the limiting inventory variance will be bounded only for stable cases.

5.1.3 The complete solution

We can also note from the above discussion and work up that we have also obtained the stochastic part of the solution, as: $x_3(k)^{stoc} \equiv \sum_{l=0}^{k-1} \beta_l \varepsilon(k-l)$, where the βs are as has been obtained above. Thus, the complete solution can be written as:

$$x_3(k) \equiv x_3(k)^{\det} + x_3(k)^{stoc} \equiv x_3(k)^{\det} + \sum_{l=0}^{k-1} \beta_l \varepsilon(k-l) \tag{51}$$

where $x_3(k)^{\det}$ is the mean response derived from the solution of the deterministic part of the LDE, and the βs in the stochastic part are as obtained above by solution of the stochastic part of the system equation, the SDE.

The above representation proves useful for simulation purposes.

We next take up the P(ID) control. We discuss only the deterministic system LDE hereafter, since the stochastic part of the solution and the limiting inventory can be obtained by methods similar to that discussed above in all cases to follow.

5.2 P(ID) control under zero lag

In this type of control an additional demand-triggered component is also added to the control thereby making it more proactive. The control initiates corrective replenishment action no sooner than a demand deviation is observed. It does not wait for an inventory deviation to take place before initiating replenishment action though it does have an inventory-triggered component also.

The replenishment control flow is given by:

$$q_3(k+1) \equiv K_3 x_3(k-1) + K_0^3 r_3(k-1) \tag{52}$$

where the first term is the inventory-triggered component and the second the demand-triggered component. Substituting for the control flow into the system equation yields:

$$x_3(k+1) \equiv x_3(k) + K_3 x_3(k-1) + K_0^3 r_3(k-1) - r_3(k+1) \tag{53}$$

which can be written as

$$x_3(k+1) - x_3(k) - K_3 x_3(k-1) \equiv K_0^3 r_3(k-1) - r_3(k+1) \tag{54}$$

We can note that the addition of the demand-triggered component has left the LHS of the LDE unaltered. Thus, the LHS Operator of the LDE remains the same and is unaffected by addition of demand-triggered components to the control. The system eqn. can hence be written in Operator form as:

$$\left[E^2 - E - K_3\right] x_3(k) \equiv K_0^3 r_3(k) - r_3(k+2) \text{ valid in } k \geq 0 \tag{55}$$

Since $r_3(k) \equiv 1, \forall k \geq 1$, the system LDE can be written as:

$$[E^2 - E - K_3]x_3(k) \equiv K_0^3 - 1 \text{ valid in } k \geq 1, \tag{56}$$

$$\text{with the ICs now as : } \{x_3(1) = -1, x_3(2) = -2\}. \tag{57}$$

The ICs for the LDE have been obtained from the system Eq. (55) above using the standard system ICs; $\{(x_3(k), r_3(k)) \equiv (0, 0), \forall k \leq 0\}$ applied to the Eq. (55) above.

Since the LHS Operator is the same as for the earlier P(I) control, the stability analysis remains the same as earlier, as also the roots of the LHS Operator for various values of the inventory-trigger parameter discussed earlier, i.e.,$\{K_3 = -1/4, -1/2, -1\}$.

The solutions are the same as given earlier in Eqs. (25) and (26).

And substituting the solution back into the O-NHE yields the value of the extra constant D as

$$D = \frac{K_0^3 - 1}{K_3}, \tag{58}$$

which hence yields the solutions for the two cases as:

$$x_3(k) \equiv (K_0^3 - 1)/K_3 + C_1\lambda_1^k + C_2\lambda_2^k \text{ for the case of distinct roots} \tag{59}$$

$$x_3(k) \equiv (K_0^3 - 1)/K_3 + (C_0 + C_1k)\lambda^k \text{ for the case of repeated roots } (K_3 = -1/4) \tag{60}$$

where the offset term is now given by $(K_0^3 - 1)/K_3$ for both cases.

The important point to note in the above solution is that the offset can be made zero by choice of the demand-trigger parameter as $K_0^3 = 1$.

And hence we can observe the enhanced response of the P(ID) control over the earlier P(I) control, in that the Offset can now be controlled by us by choice of K_0^3.

In fact, we can also achieve a (+)ve value of the offset by choosing $K_0^3 \geq 1$.

We can now obtain the full solutions for the three cases above, using the LDE ICs $\{x_3(1) = -1, x_3(2) = -2\}$. We take $K_0^3 = 1$ to be able to obtain zero offset. Hence, we have:

$$\text{For } K_3 = -1/4 \text{ (the repeated roots case)} : x_3(k) \equiv (4 - 6k)(1/2)^k, k \geq 1 \tag{61}$$

$$\text{For } K_3 = -1/2 : x_3(k) \equiv \left(2\sqrt{5}\right)\text{Cos}(k\pi/4 - \phi)\left(1/\sqrt{2}\right)^k, \tan\phi = 2, k \geq 1 \tag{62}$$

$$\text{For } K_3 = -1 \text{ (the marginal stability case)} : x_3(k) \equiv 2\text{Cos}((k+1)\pi/3), k \geq 1 \tag{63}$$

The solution curves are plotted in **Figure 3**, from which we can see that the response in all cases has zero offset.

We can similarly extend the modeling and analysis to PI(I), PID(I), and MA(ID) controls.

We could also have different types of input disturbances as indicated earlier in Section 3.1. Additionally, the third dimension of our analysis could be to have non-zero lags, i.e., lags of one, two periods, and so on. Cases with non-zero and higher lags will result in higher order system LDEs, and the LHS Operator would be of a higher order.

Further details of the above can be obtained in [1–9].

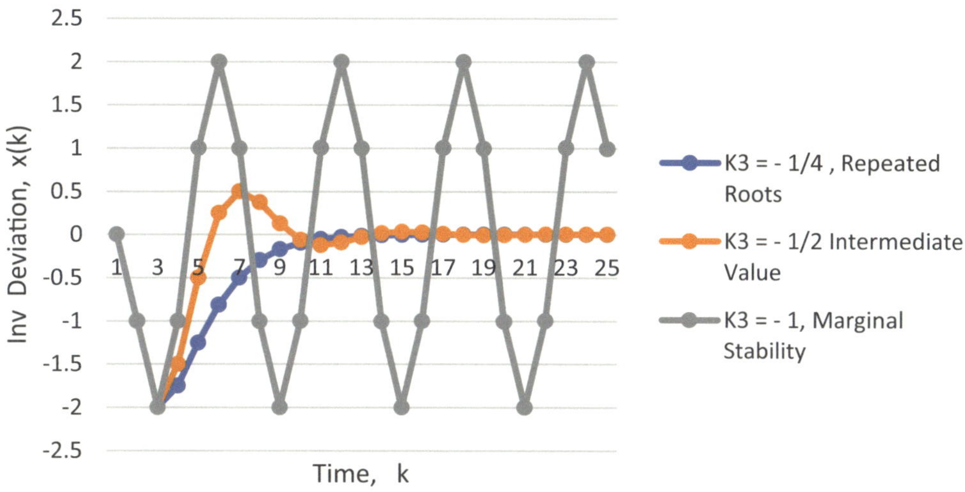

Figure 3.
P(ID) control.

6. Controls for multi-stage supply chains

We look at the serial supply chain system as given in **Figure 1**.

We can see from **Figure 1** that the immediately succeeding downstream stage in a supply chain will provide the "demand perturbation" for the immediately preceding stage. Thus, the demand perturbation at the warehouse at the downstream end will successively be felt up the chain. And the single-stage analysis described above can be used in turn for each stage of the chain.

For non-serial supply chains, the arguments are similar and single-stage analysis can be used as described above.

Details of some of these analyses can be found in [1–9].

7. Conclusion

This chapter has presented the application of control concepts to the control of supply chains. The state variables have been taken to be the inventory levels, while the control variables are the replenishment flows into the various stages of the system. The conventional P, PI, PID controls have been discussed, as also some newer forms of control which are especially applicable to supply chains and warehouses. The performance of P(I) and P(ID) controls have been derived in detail, and their performance analyzed.

A significant feature of this chapter is that the conventional block diagrams and transfer functions of conventional control theory have not been used. Rather direct Operator Methods have been used to good advantage to solve the system equations.

Author details

Kannan Nilakantan
Mathematical Modelling Group, Operations Group, Institute of Management Technology, Nagpur, India

*Address all correspondence to: nilakanthan@gmail.com

IntechOpen

References

[1] Nilakantan, Kannan; Enhancing Supply Chain performance with improved Order-control Policies, *International Journal of System Science* Sep2010, Vol 41-9, p 1099-1113, Taylor and Francis.

[2] Nilakantan, Kannan; Replenishment Policies for Warehouse Systems under Cyclic demand, *International Journal of Business Performance and Supply Chain Modelling*, 2013,Inderscience, Vol 5-2, pp 148-176.

[3] Nilakantan, Kannan; Supply Chains subject to Demand Shocks, *International Journal of Logistics Systems and Management,* 2015, Inderscience, Vol 21-2, pp133-159.

[4] Nilakantan, Kannan; Supply Chains under Demands with Trend Components, and Robust Controls, *Annals of Management Science*, International Centre for Business and Management Excellence, Franklin, Tennessee, USA.Vol 3-2, Dec 2014, pp 27-64.

[5] Nilakantan, Kannan; Warehouse Control Systems for General Forms of Demand, *International Journal of Operations and Quantitative Management*, 2014, INFOMS, USA, Vol 20–4, pp 273–299,Dec 2014.

[6] Nilakantan, Kannan; Replenishment systems for responsive supply chains under dynamic and sudden lead-time disturbances, *International Journal of Systems Science-Operations and Logistics*, 2019, Elsevier, Vol 6-4, pp 320-345.

[7] Nilakantan, Kannan; Supply Chain Resilience to Sudden and Simultaneous Lead-time and Demand Disturbances *International Journal of Supply Chain and Operations Resilience,* 2019, Inderscience (USA), Vol 3 - 4, pp. 292-329.

[8] Nilakantan Kannan, Responsiveness and Recovery Performance Analysis, and Replenishment System Design in Supply Chains under Large Replenishment Lead-times and Uncertain Demand, I*nternational Journal of Supply Chain Management 2021,(UK),* Vol 10-3, pp 1-17.

[9] Nilakantan Kannan The Ripple Effect in a Supply Chain: A Sudden Demand Increase and with a Sinusoidal Component, International Journal of Operations and Quantitative Management March *2021*, Vol 27-1, pp. 1-37.

[10] Kelley, W. G., & Peterson, A. C., *Difference Equations*, 2/e, Academic Press, Elsevier, USA.

[11] Gourieroux, G, & Monfort, A., Time Series and Dynamic Models, Cambridge University Press,(1990), Cambridge, UK.

Chapter 4

The Fundamental of TCP Techniques

Pritee Nivrutti Hulule

Abstract

Strategies for prioritizing test cases plan test cases to reduce the cost of retrospective testing and to enhance a specific objective function. Test cases are prioritized as those most important test cases under certain conditions are made before the re-examination process. There are many strategies available in the literature that focus on achieving various pre-test testing objectives and thus reduce their cost. In addition, inspectors often select a few well-known strategies for prioritizing trial cases. The main reason behind the lack of guidelines for the selection of TCP strategies. Therefore, this part of the study introduces the novel approach to TCP strategic planning using the ambiguous concept to support the effective selection of experimental strategies to prioritize experimental cases. This function is an extension of the already selected selection schemes for the prioritization of probation cases.

Keywords: test case prioritization (TCP), final total effort (FTE), average effort (AE)

1. Introduction

In testing, part of Regression testing in the maintenance phase is the process of retesting the updated software to ensure that new errors have not been introduced into earlier validated code. In addition, the regression tests should take as little time as possible to perform a few test cases as possible. Due to its costly nature, several techniques in the literature focus on costs.

These are:

i. Re-run everything

ii. Minimization/reduction of the test case

iii. Selection of the test case

iv. Prioritization of the test case

v. Hybrid approach.

This document focuses on the techniques of prioritization of the test case. Testers may now want to increase code coverage in test software at a faster pace, increase or improve their reliability in software reliability in less time, or increase the speed at which test suites detect failures at that moment. System during the regression tests. The main problems with code-based prioritization techniques are that they focus only on the number of errors detected and therefore treat all failures in the same way [1–4].

2. Test cases prioritization

Testing software or applications is the most important part of the "Software Development Life Cycle" (SDLC). It plays a very important role in the quality and performance of the software and ensures that the final product is as per the client's requirement. Placement Priority is the expansion of software testing, which is used to determine the "critical test cases". Software testing is done to detect bugs and errors in its operation, depending on how the performance and quality of the software are continuously improved.

These preliminary test cases are determined by a variety of factors depending on the need for the software, which is provided for test cases to perform other processes. By prioritizing test cases, (testers and developers can reduce the time and cost of the software testing phase and ensure that the product delivered is of different quality) [3–6].

2.1 What is test prioritization?

The term "test prioritization" refers to the accountable and difficult part of testing that allows assessors to "manage risk, plan tests, estimate cost, and analyze" which tests will work in the context of a particular project. This process is known as Test Case Prioritization, which is the process of 'prioritizing and planning' test cases. These methods are used to run test cases that are very important to reduce time, cost, and effort during the software testing phase.

In addition, the prioritization of the test case helps in regression testing and improves its effectiveness Developers of forensic trial software can get help to fix bugs earlier than possible otherwise. In addition, to determine the prioritization of test cases, different factors are determined according to the need of the software.

In this way, inspectors can easily run test cases, which have a very high value and provide errors with previous defects. Also, the improved error detection rate during the test phase allows for faster response of the test system.

The major issues of code-based prioritization techniques are that they focus only on the number of faults detected and hence treats all faults equally. In practice, all faults cannot be treated the same. Therefore, there may be situations where the presence of an error is less important but its coverage of the requirements is important. The prioritization of needs-based assessment addresses those issues by providing the most relevant case-based assessment of service-based services. In addition, Testing does not guarantee error-free software is a process of verifying software compared to user descriptions and their requirements major problem with the specifics of Test Case Prioritization is based on specificity and that there is no effective way to measure the performance of selected Test Suits [3].

Strategies for the promotion of probation cases make probationary cases subject to certain conditions. Prioritization of test cases can serve a variety of purposes.

The purpose of these priorities increases the likelihood that they will meet a particular goal more closely than they would otherwise have done at random. The Test Case Prioritization problem was officially described by Rothaermel and she learned nine TCP techniques. Among them, four relied on coding and two relied on early detection of errors. This was compared with no priorities and random prioritization strategies. The right category can only happen in books that are likely to 'benefit. Tests show an early detection of error with the greedy algorithm and additional greed in the code detection process. To measure the purpose of the experiment Rotermel defined metrics, the APFD rated an average of% of errors found over a fraction of the test suit (%) made. Its values are between 0 and 100 and the higher the value the better the error detection [3].

2.2 Intelligence techniques

2.2.1 Uncertainty and metaheuristics

The essence of this study is that it can be used in both the White-box test areas or the black box test environment. This approach aims to redesign test cases according to the extent of the alleged violation from the source code. The process of prioritizing probation cases using the absurd concept to improve the performance of a given assessment suit in violation of the evidence and further prioritization of probation cases. Anwar et al. conducted a study comparing various experimental precautionary measures and used the ambiguous notion of making good use of an experimental suit which is why the testing process backfired. Risk assessment of software needs is a major factor in improving software quality. Propose a new risk assessment approach using a sophisticated professional program to improve the effectiveness of TCP in the review process. All of these studies demonstrate the development of strategic priorities for testing. As time changes, the nature of these subjects changes. The proposed approach is a systematic study that minimizes theoretical aspects and drives research into a practical context. Although many strategies have been proposed many strategies are limited to code-based methods and focus on detecting a high number of errors. According to a study conducted by Catal (2013) on Test Case Prioritization Techniques, the highest number of strategies proposed so far is Coverage (code) based (40%) and minimal value is given to known costs (2%) and distribution-based (2%) strategies.

2.3 Benefits of proposed methodology

The major issues of code-based prioritization techniques are that they focus only on the number of faults detected and hence treats all faults equally. That is removed from the system.

There may be some cases where the existence of fault is not so important but its requirement coverage is that is performed in the proposed technique.

This part of the study is an extension of the selected selection schema and provides a Model of Priority Testing Priority Strategies based on three factors:

i. Service delivery,

ii. Efforts

iii. Complexity

In selecting the schema identification of project features/features that need to be done to identify TCP strategies that include high project attributes and therefore requirements. Selection Schema assumes that the required tools are the same. Once the strategy has been identified the next step is to differentiate those strategies based on high integration and a small experimental effort again, with difficulty. To calculate the effort this study uses the same basis as used by Krishnamoorthi. Therefore, the experimental effort represents the average number of test cases required for a specific functional evaluation process. Methodology plays a role in the evaluation of experimental efforts. According to research, the difficulty can be taken from a scale of 1–10 which is usually defined by the engineer and analyst. Therefore, this part of the study achieves high coverage with minimal difficulty and experimental efforts [2, 7].

3. Phases of TCP

Stage-1 uses the priority matrix proposed in 'the Selection Schema'.

Stage-2 experiences the difficulty of a particular method compared to the availability of its requirements.

TCP strategy tension is measured on a scale of (1–10) based on research. Calculating the effort of a particular method above number 2 is used. According to the formulas, the effort will come on a scale of (0–1). In this way, we will get used to multiplying by 10 so that we consider the same amount as coverage and weight. In the final stage, the classification is done with the Fuzzy Rule-based system.

The output of this program is sorted into the following upcoming sets:" Select, Medium and, Discard". Input variables, as well as output variables, can take values between (1–10). In this case study, triangular membership functions are used for mapping random and flexible input sets during fuzzification as well as for making dynamic output and complex sets during defuzzification. The input variables are written in three non-linear sets each: Low, Middle, and Top various purposes. The priority of the test case can be explained.

Given: In the test, suit provided S of the given system (X; XS, set of S) permissions;

f, function from XS to real numbers. Problem: Get S'ϵ XS to (6S ") (S"ϵXS) (S's's ') [f (S') ≥ f (S ")].

In this definition, 'XS' is a collection of existing combinations to prioritize test cases for test 'S', and f is an objective function. For example, testers may wish to increase code coverage in software under test at a faster rate, increase or improve their confidence in software reliability in the short term, or increase the rate at which test suits find errors in that system during deferred testing.

The Test case Prioritization problem was officially described by Rothermel and he learned the nine TCP techniques discussed. Four of them were based on coding and two were based on the early detection rate. This was compared with no priorities and random prioritization strategies. The appropriate category is only possible in the text that is almost impossible to achieve. Tests show an early detection of error with the greedy algorithm and additional greed in the code detection process. To measure the purpose of the experiment Rothermel explained the metric, the APFD measuring an average of % of errors found over a fraction of the test suit (%) made. Its values are between 0 and 100 and the higher the value the better the error detection.

Think of a test suit T with several test cases; F is a set.

of m faults detected with test T T. TFi is the first test case in T'(one of T's orders) indicating error i. Thereafter T "s APFD is defined by the following equation [1, 4, 5, 8, 9]:

$$Average\ Percentage\ of\ Fault\ Detected = 1 - \left(\frac{TF1 + TF2 + \cdots TFm}{nm}\right) + \left(\frac{1}{2n}\right) \quad (1)$$

4. Methods

This proposed law-based program has a total of "17 rules" and is reviewed frequently based on expert knowledge. These rules are based on the following experts.

4.1 Motivation and contribution

In previous research, the researcher focuses on factors that I use, but they are not focused on the most important part 'time'. I work on that, and also how many testers are using the system, that tester priority is" low, medium or high" that things also captured and it generates the graph on basis of that.

4.2 Research findings

1. With the selection of any process the most important factor is the installation of project signs and therefore the installation of requirements.

2. Efforts are determined and calculated once the requirements have been met. Therefore, strategies should be chosen with "high coverage and few attempts" to achieve those features.

3. Complexity also plays an important role because only complexity determines the effort of a particular method (**Table 1**).

4.3 Method analysis

There Are Four Factors:

i. Requirement coverage

ii. Efforts

iii. Complexity

iv. Time

4.3.1 Input (three inputs)

1. Relevance of selected TCP Techniques based on maximum requirement coverage.

Rule#	Rule hypothesis	Class (rule outcome)
1	If (rel = low && comp + high && effort = low)	Discard
2	If (rel = low && comp + high && effort = medium)	Discard
3	If (rel = low && comp = high & & effort + high)	Discard
4	If (rel = low && comp = high && effort = high)	Discard
5	If (rel = high && comp = medium && effort = medium)	Select
6	If (rel = high && comp = low && effort = medium)	Select
7	If (rel = high && comp = low && effort = low)	Select
8	If (rel = high && comp = medium && effort = low)	Select
9	If(rel = high && comp + high && effort = low)	Select
10	If (rel = medium && comp + high && effort = low)	Moderate
11	If (rel = medium && comp + high && effort = medium)	Moderate
12	If (rel = medium && comp + high && effort = high)	Discard
13	If (rel = medium && comp = high && effort = high)	Discard
14	If (rel = medium && comp + high && effort = medium)	Moderate
15	If (rel = low && comp = high && effort + medium)	Discard
16	If (rel == medium && comp == high & & effort == medium)	Discard
17	If (rel == high && comp == medium && effort == high)	Discard

Table 1.
Rule base for fuzzy based selection of TCP techniques.

2. The complexity of selected TCP techniques

3. Average Effort (AE)

4. Time to execute

Output: Final class: TCP Techniques [2].

4.3.2 Begin

1. Identify input variables (linguistic variables) i.e., "relevance, AE, Complexity" (initialization).

2. Mapping of fuzzy sets to input variables by constructing the membership functions (initialization).

3. Formation of rules to create the rule base (initialization).

4. Conversion of input data (fuzzification).

5. Assessment of available rules in the rule base (inference).

6. Merge all the results achieved from available rules (inference).

7. Mapping of output data (defuzzification) [10].

4.3.3 Architecture

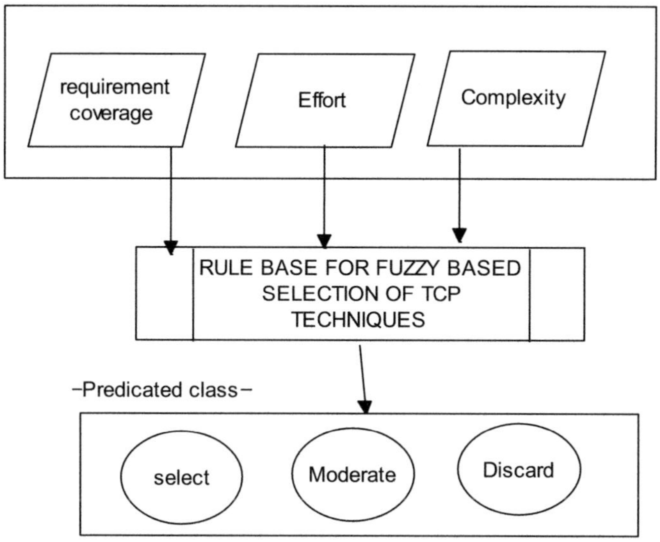

relevant project 'attributes/features' are done to identify TCP techniques covering maximum project attributes consequently requirements.

The various stages of the proposed approach are as follows:

Stage-1 Identifying project features in terms of relevance and hence the coverage of requirements.

Stage-2 Identify the complexity of testing techniques.

Stage-3 calculating testing effort.

Stage-4 classifies TCP techniques using fuzzy inference.

Stage 5. Time to execute each technique. Selection of any technique most important factor is time to perform an execution [2, 11].

4.4 Technical feasibility

Technical feasibility deals with the study of performance and numerous constraints like availableness of "resources, technology", the risk concerning development that might affect the capability to attain an adequate system. It identifies if the technology used is companionable or not with the recent system.

Following are some technical issues are:

- The system needed a large set of 'MySQL' database connectivity.
- This system technically supports different development tools like Eclipse, Net beans, etc.

4.5 Economic feasibility

Economic analysis is the generally used methodology for estimating the efficiency of a recent system. The procedure of cost analysis is to establish the advantages and savings that are looking ahead from a system and evaluate them with their costs.

Time-Based: On a single click management can create any report.

Cost-Based: There is no need for any type of training to use the software or tools. For managing this tool there is no need for any investment. The software is freeware so there is a minimal cost [2].

4.6 Operational feasibility

Operational feasibility is the activity of however well a projected system solves the issues. It receives the benefits of the opportunities known throughout the scope definition and the way it satisfies the needs known within the requirements analysis part of system implementation. If the user is aware of the all technicalities used for the system and the user having in detail knowledge about the system, then there are no difficulties at the time of implementing the system. Therefore, it's assumed that the user will not face any downside once handling the system implantation. Getting acceptance from users is the major difficulty for any developer at the time of developing any software tool. This system does not have any major problems. The small problem is that some time system gets slow due to the server and makes us wait for results. But it will automatically generate results fast. This is not a permanent problem of this system.

Case Studies for Development and Implementation:

1. With the selection of any process the most important factor is the installation of project signs and therefore the installation of requirements.

2. Efforts are determined and calculated once the requirements have been met. Therefore, strategies should be chosen with high coverage and few attempts to achieve those features.

3. Complexity also plays an important role because only complexity determines the effort of a particular method.

4. Time: Time to use TCP.

5. Mathematical model

A] Mapping Diagram.

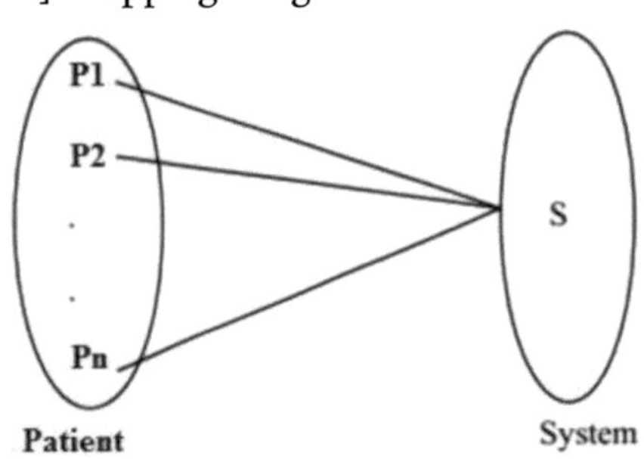

Where,
P1, P2 .... Pn = Tester.
S = System.
B] **Set Theory.**
S = {s, e, X, Y, ϕ}.
Where,
s = Start of the program.

1. Authentication

L = Login, UN = User name, PWD = Password.

To access the facilities of system, TCP Classification.

X = Input of the program.

2. Identify INPUT as

Input should be 4 factors.

X = {S1, S2 … … Sn}.

Where,

S1, S2 … … Sn = No. of Factored selected by Patients.

3. Identify Process P as

P = {DF}.

Where,

DP = TCP Classification rules

4. Identify Output Y as

Y = {Bc1 … …Bcn}.

Where,

Bc1 … …Bcn = Select, discard, moderate class.

According to condition selected factor parameter TC will classify in one Class φ = Success or failure condition of the system [4].

6. Failures

1. A huge database can lead to more time-consuming to get the information.

2. Hardware failure.

3. Software failure.

7. Success

1. Search the required information from available rules.

2. The user gets results very fast according to their needs.

3. The mathematical model above is NP-Complete.

8. Our project is NP-complete

Our project goes into NP-Finish because at some point it will give the result. With the resolution problem, so that it will provide a solution to the problem during the polynomial period. A collection of all decision-making problems the solution to which can be provided in the polynomial period. Functional requirement.

8.1 User

- Tester login to the system.
- Update profile
- Add tcp info for classification
- View predicted class of tcp with four parameters

    i. Requirement coverage,

    ii. Efforts

    iii. Complexity

    iv. Time. View own TCP.

8.2 Admin

- Admin will log in to the system.
- Admin activates tester.
- Admin can view all TCP Classification classes.
- Admin can view all TCP Classification graphs with 3 categorize.

9. Software quality attributes

9.1 Capacity

The capacity of a project according to data is very less.

9.2 Availability

All functionality will work properly.

9.3 Reliability

The system is reliable for classifying large data.

9.4 Security

User when login to the system that time users Mail Id and password match accurately.

9.5 DFD diagram

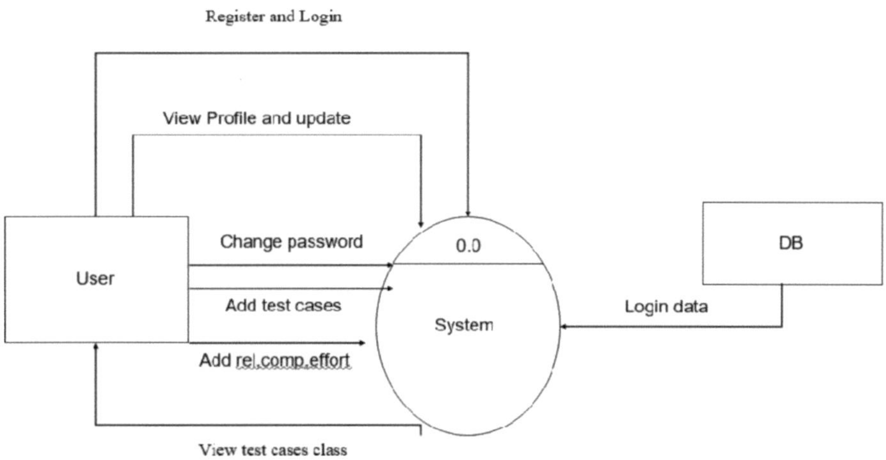

Fig-Data Flow Diagram.

System Use Case Diagram:

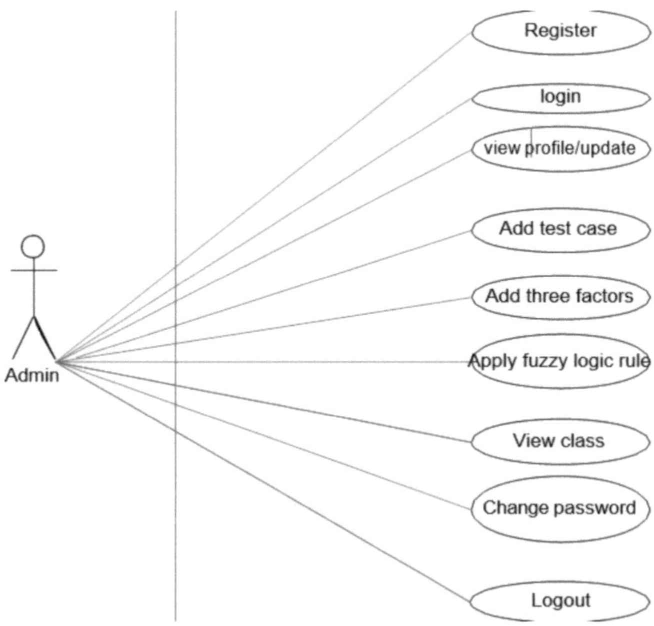

Fig-Use Case Diagram.

Class Diagram:

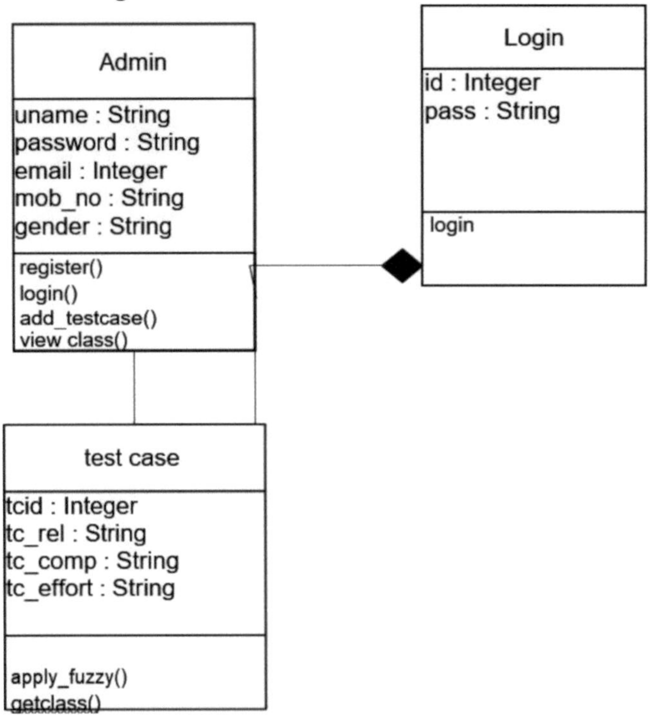

Fig-Class Diagram.

10. Architecture Modeling

Component Diagram

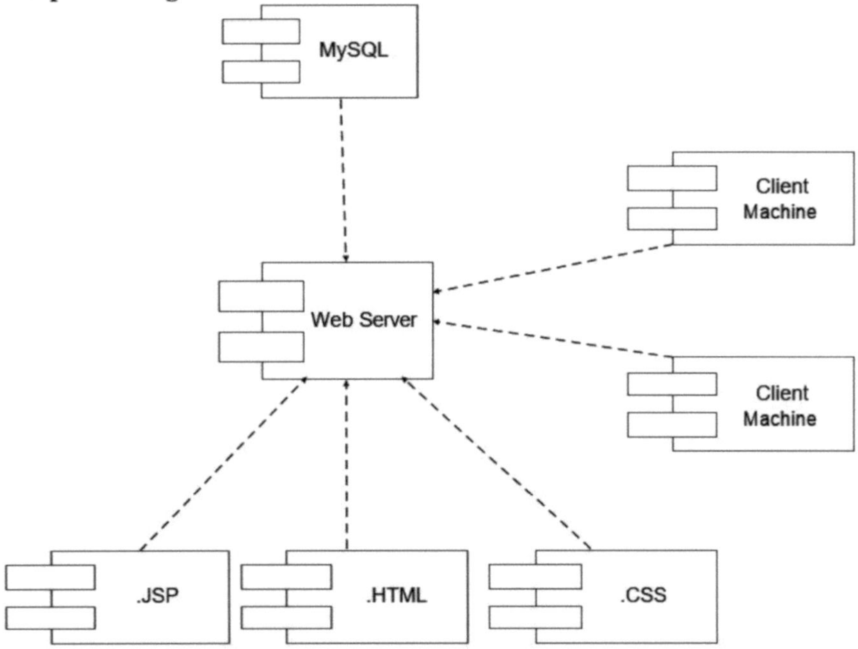

Deployment Diagram:

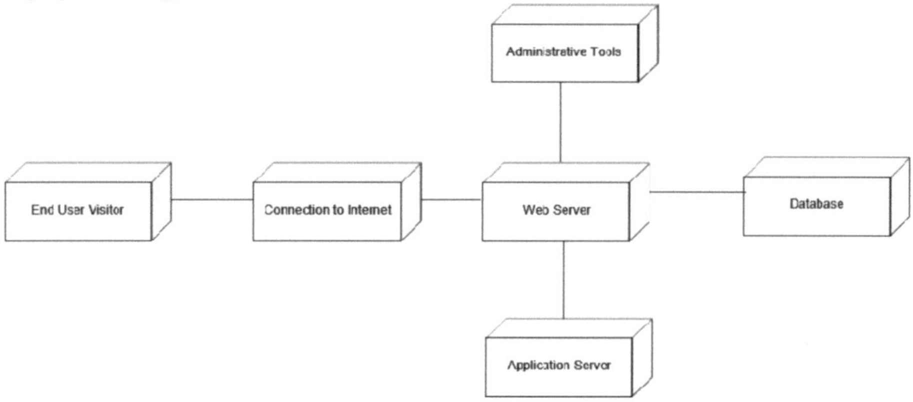

11. The design process for quality software

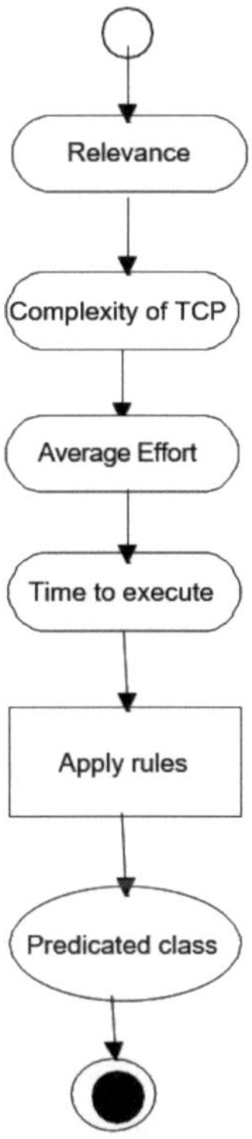

11.1 Implementation approach

Describe the overall test method that will be used to evaluate the product of the project.

There are many ways such as:

- Black Box check
- White Box test

Here we have used the "Black Box test" method. In Black Box Testing we simply provide input to the system and test its output regardless of how the system processes it.

11.2 Passport test or test failure terms

Where the actual results and expectations are the same the test will be passed. When the actual results and expectations are different the test will fail.

Entry method: as soon as we have a need, we can start testing.

Exit method: when the disturbance level falls below a certain level, we can stop the test.

11.3 Implementation with screen output

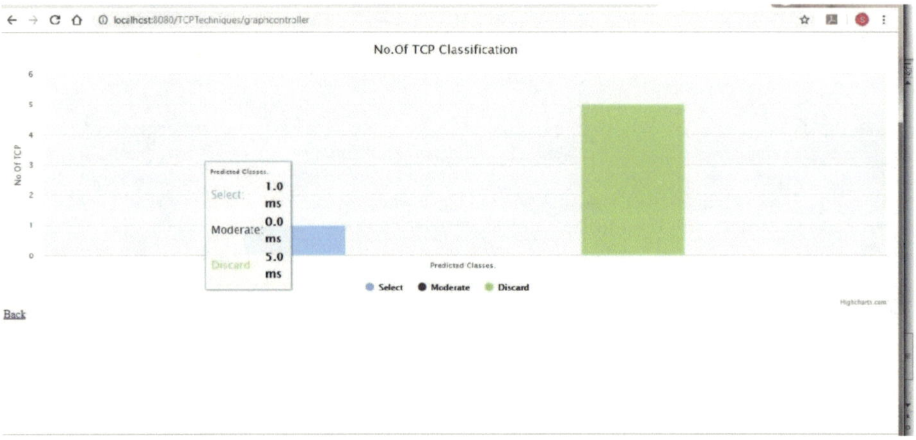

Present the system in this research we have proposed a novel-based technique for the classification of TCP techniques using Fuzzy Logic. "This work is an extension of the already proposed selection schema for test case prioritization techniques". with the help of tester login and that tester priority of use the factor, it can generate the graph and also, it how much time required for the execution of the test case it calculates. Model for selection of test case prioritization technique based on 4f factors:

i. Requirement coverage

ii. Efforts

iii. Complexity

iv. Time

12. Future scope

"The system will work for other projects testing like ERP.TCP techniques will enhance performance by using another solution to prioritize test cases".

Author details

Pritee Nivrutti Hulule
Computing Engineering, Bharati Vidyapeeth (Deemed to be University) College of Engineering, Pune, India

*Address all correspondence to: hululepritee9@gmail.com

References

[1] Duggal G, Suri B. Understanding Regression Testing Techniques. COIT; 2008

[2] Sujata KK, Purohit GN. A schema support for selection of test case prioritization techniques. In: Fifth International Conference on Advanced Computing & Communication Technologies (ACCT '15). 2015. pp. 547-551

[3] Vegas S, Basili V. A characterization schema for software testing techniques. Empirical Software Engineering. 2005; **10**(4):437-466

[4] Yoo S, Harman M. Regression testing minimization, selection, and prioritization: a survey. Software Testing, Verification, and Reliability. 2012;**22**:67-120

[5] Tyagi M, Malhotra S. An approach for test case prioritization based on three factors. International Journal of Information Technology and Computer Science. 2015;**4**:79-86

[6] Kumar V, Sujata K, Kumar M. Test case prioritization using fault severity. International Journal of Computer Science and Technology (IJCST). 2010; **1**(1):67-71

[7] Chen GY-H, Wang P-Q. Test case prioritization in specification-based environment. Journal of Software. 2014; **9**(8):205-2064

[8] Elbaum S, Malishevsky A, Rothermel G. Test case prioritization: a family of empirical studies. IEEE Transactions on Software Engineering. 2002;**28**:159-182

[9] Miranda B, Cruciani E. 2018 Copyright held by the owner/ author(s), FAST approaches to scalable similarity-based test case prioritization. 2018

[10] Silva D, Rabelo R, Campanhã M, Neto PS, Oliveira PA, Britto R. A hybrid approach for test case prioritization and selection. In: IEEE Congress on Evolutionary Computation (CEC). 2016. pp. 4508-4515

[11] Alakeel AM. Using fuzzy logic in test case prioritization for regression testing programs with assertions. The Scientific World Journal. 2014;**2014**: Article ID-316014

Section 2

Experimental Analysis and TCP Technique

Chapter 5

On Parametrizations of State Feedbacks and Static Output Feedbacks and Their Applications

Yossi Peretz

Abstract

In this chapter, we provide an explicit free parametrization of all the stabilizing static state feedbacks for continuous-time Linear-Time-Invariant (LTI) systems, which are given in their state-space representation. The parametrization of the set of all the stabilizing static output feedbacks is next derived by imposing a linear constraint on the stabilizing static state feedbacks of a related system. The parametrizations are utilized for optimal control problems and for pole-placement and exact pole-assignment problems.

Keywords: control systems, continuous-time systems, state-space representation, feedback stabilization, static state feedback, static output feedback, Lyapunov equation, parametrization, optimization, optimal control, H_∞-control, H_2-control, linear-quadratic regulators, pole assignment, pole placement, robust control

1. Introduction

The solution of the problem of stabilizing static output feedback (SOF) has a great practical importance, for several reasons: they are simple, cheap, reliable, and their implementation is simple and direct. Since in practical applications, full-state measurements are not always available, the application of stabilizing state feedback (SF) is not always possible. Obviously, in practical applications, the entries of the needed SOFs are bounded with bounds known in advance, but unfortunately, the problem of SOFs with interval constrained entries is NP-hard (see [1, 2]). Exact pole assignment and simultaneous stabilization via SOF or stabilization via structured SOFs are also NP-hard problems (see [2, 3] resp.). These problems become even harder when optimal SOFs are sought, when the optimality notions can be the sparsity (see [4]) of the controller (e.g., for reliability purposes of networked control systems (NCSs)), the cost or energy consumption of the controller (which are related to various norm-bounds on the controller), the H_∞-norm, the H_2-norm or the linear-quadratic regulator (LQR) functional of the closed loop. The practical meaning of the NP-hardness of the aforementioned problems is that the problems cannot be formulated as convex problems (e.g., through LMIs or SDPs) and cannot have any efficient algorithms (under the widespread belief that P $\neq$ NP). Thus, one has to compromise the exactness (which might affect the feasibility of the solution) or the optimality of the solution. Therefore, one has to utilize the specific structure of the given problem, in order to describe effectively the set of all feasible solutions, by reducing the number of variables and constraints to the minimum, for the purpose of increasing the efficiency and accuracy of the available algorithms. This is the aim of the proposed method.

Several formulations and related algorithms were introduced in the literature for the constrained SOF and other control hard problems. The iterated linear matrix inequalities (ILMI), bilinear matrix inequalities (BMI), and semi-definite programming (SDP) approaches for the constrained SOF problem, for the simultaneous stabilizing SOF problem, and for the robust control via SOF (with related algorithms) were studied in: [5–11]. The problem of pole placement via SOF and the problem of robust pole placement via Static Feedback were studied in: [12, 13]. In [14, 15], the method of alternating projections was utilized to solve the problems of rank minimization and pole placement via SOFs, respectively. The probabilistic and randomized methods for the constrained SOF problem and robust stabilization via SOFs (among other hard problems) were discussed in [16–19]. In [20], the problem of minimal-gain SOF was solved efficiently by the randomized method. A non-smooth analysis approach for H_∞ synthesis and for the SOF problem is given in [21, 22], respectively. A MATLAB® library for multiobjective robust control problems based on the non-smooth analysis approach was introduced in [23]. All these references (and many more references not brought here) show the significance of the constrained SOF problem to control applications.

Many problems can be reduced to the SOF constrained problem, including the reduction of the minimal-degree dynamic-feedback problem and robust or decentralized stability via static-feedback, reduced-order H_∞ filter problem, global minimization of LQR functional via SOF, and the design problem of optimal PID controllers (see [2, 10, 24–27] respectively). It is worth mentioning [28], where the alternating direction method of multipliers was utilized to alternate between optimizing the sparsity of the state feedback matrix and optimizing the closed-loop H_2-norm, where the sparsity measure was introduced as a penalty term, without any pre-assumed knowledge about the sparsity structure of the controller. The method of augmented Lagrangian for optimal structured static-feedbacks was considered in [29], where it is assumed that the structure is known in advance (otherwise, one should solve a combinatorial problem). The computation overhead of all the aforementioned methods can be reduced significantly, if good parametrization of all the SOFs of the given system could be found, where a parametrization can be called "good" if it takes into account the structure of the given specific system and if it well separates between free and dependent parameters, thus resulting in a minimal set of nonlinear nonconvex inequalities/equations needed to be solved.

In [30], a parametrization of all the SFs and SOFs of Linear-Time-Invariant (LTI) continuous-time systems is achieved by using a characterization of all the (marginally) stable matrices as dissipative Hamiltonian matrices, leading to a highly performance sequential semi-definite programming algorithm for the minimal-gain SOF problem. The proposed method there can be applied also to LTI discrete-time systems by adding semi-definite conditions for placing the closed-loop eigenvalues in the unit disk. A new parametrization for SOF control of linear parameter-varying (LPV) discrete-time systems, with guaranteed ℓ_2-gain performance, is provided in [31]. The parametrization there is given in terms of an infinite set of LMIs that becomes finite, if some structure on the parameter-dependent matrices is assumed (e.g., an affine dependency). The H_2-norm guaranteed-performance SOF control for hidden Markov jump linear systems (HMJLS) is studied in [32], where the SOFs are parameterized via convex optimization with LMI constraints, under the assumptions of full-rank sensor matrices and an efficient and accurate Markov chain state estimator. In [33], an iterative LMI algorithm is proposed for the SOF problem for LTI continuous-time negative-imaginary (NI) systems with given H_∞ norm-bound on the closed loop, based on decoupling the dependencies between the SOF and the Lyapunov certificate matrix.

When solving an optimization problem, it is important to have a convenient parametrization for the set of feasible solutions. Otherwise, one needs to use the probability method (i.e., the "generate and check" method), which is seriously doomed to the "curse of dimensionality" (see [16]). In [13], a closed form of all the stabilizing state feedbacks is proved (up to a set of measure 0), for the purpose of exact pole assignment, when the location errors are optimized by lowering the condition number of the similarity matrix, and the controller performance is optimized by minimizing its Frobenius norm. The parametrization in [13] is based on the assumptions that the input-to-state matrix B has a full rank and at least one real state feedback leading to diagonalizable closed-loop matrix exists, where a necessary condition for the existence of such feedback is that the multiplicity of any assigned eigenvalue is less than or equal to $rank(B)$. In this context, it is worth mentioning [34] in which a parametrization of all the exact pole-assignment state feedbacks is given, under the assumption that the set of needed closed-loop poles should contain sufficient number of real eigenvalues (which make no problem if the problem of pole placement is of concern, where it is generally assumed that the region is symmetric with respect to the real axis and contains a real-axis segment with its neighborhood). The results of [34] and of the current chapter are based on a controllability recursive structure that was discovered in [35].

In this chapter, using the aforementioned controllability recursive structure, we introduce a parametrization of the set of all stabilizing SOFs for continuous-time LTI systems with no other assumptions on the given system (for discrete-time LTI systems, the parametrization is much more involved and will be treated in a future work). As opposed to the notable works [36–38], where for the parametrization one still needs to solve some LMIs in order to get the Lyapunov matrix, here we give an explicit recursive formula for the Lyapunov matrix and for the feedback in the case of SF, and a constrained form for the Lyapunov matrix and for the feedback in the case of SOF.

The rest of the chapter goes as follows:

In Section 2, we set notions and give some basic useful lemmas, and in Section 3, we introduce the parametrization of the set of all stabilizing static-state feedbacks for LTI continuous-time systems. In Section 4, we introduce the constrained parametrization of the set of all stabilizing SOFs for LTI continuous-time systems. The effectiveness of the method is shown on a real-life system. Section 5 is based on [34] and is devoted to the problem of exact pole assignment by SF, for LTI continuous-time or discrete-time systems. The effectiveness of the method is shown on a real-life system. Finally, in Section 6, we conclude with some remarks and intentions for a future work.

2. Preliminaries

By $\mathbb{C}$ we denote the complex field and by $\mathbb{C}_-$ the open left half-plane. For $z \in \mathbb{C}$ we denote by $\Re(z)$ its real part, while by $\Im(z)$ we denote its imaginary part. For a square matrix Z, we denote by $\sigma(Z)$ the spectrum of Z. For a $\mathbb{R}^{p\times q}$ matrix Z, we denote by Z^T its transpose, and by $z_{i,j}$ or by $Z_{i,j}$, its (i,j)'th element or block element. A square matrix Z in the continuous-time context (in the discrete-time context) is said to be (asymptotically) stable, if any eigenvalue $\lambda \in \sigma(Z)$ satisfies $\Re(\lambda) < 0$, i.e., $\lambda \in \mathbb{C}_-$ (satisfies $|\lambda| < 1$, i.e. $\lambda \in \mathbb{D}_-$ where $\mathbb{D}_-$ is the open unit disk).

Consider a continuous-time system in the form:

$$\begin{cases} \frac{d}{dt}x(t) = Ax(t) + Bu(t) \\ y(t) = Cx(t) \end{cases} \tag{1}$$

where $A \in \mathbb{R}^{n\times n}, B \in \mathbb{R}^{n\times m}, C \in \mathbb{R}^{r\times n}$, and x, u, y are the state, the input, and the measurement, respectively. Assuming that the state x is fully accessible and fully available for feedback, we define $u = -K^{(0)}x$ to be the state feedback (SF). When the state is not fully accessible or not fully available for feedback but the measurement y is available for feedback, we define $u = -Ky$ to be the static output feedback (SOF). The problems that we consider here are the following:

- (SF-PAR): How can one parameterize the set of all $K^{(0)} \in \mathbb{R}^{m\times n}$ such that the closed-loop $A - BK^{(0)}$ is stable, and what is the best parametrization (in terms of minimal number of parameters and minimal set of constraints)?

- (SOF-PAR): How can one parameterize the set of all $K \in \mathbb{R}^{m\times r}$ such that the closed-loop $A - BKC$ is stable, and what is the best parametrization?

The parameterizations will be used for achieving other goals and performance keys for the system, other than stability, which is the feasibility defining basic key.

A square matrix Z is said to be non-negative (denoted as $Z \geq 0$) if $Z^T = Z$ and $v^T Zv \geq 0$ for any vector v. A non-negative matrix Z is said to be strictly non-negative (denoted as $Z > 0$) if $v^T Zv > 0$ for any vector $v \neq 0$. For two square matrices Z, W, we would write $Z \geq W$ $(Z > W)$ if $Z - W \geq 0$ (respectively, if $Z - W > 0$). For a matrix $Z \in \mathbb{R}^{p\times q}$, we denote by Z^+ the Moore-Penrose pseudo-inverse (see [39, 40] for definition and properties). By L_Z, R_Z we denote the orthogonal projections $I_q - Z^+Z$ and $I_p - ZZ^+$, respectively, where I_s denotes the identity matrix of size $s \times s$. Note that Z^+Z and ZZ^+ (as well as L_Z and R_Z) are symmetric and orthogonally diagonalizable with eigenvalues from $\{0, 1\}$. By *diag* and *bdiag* we denote diagonal and block-diagonal matrices, respectively.

A system triplet (A, B, C) is SOF stabilizable (or just stabilizable) if and only if there exist K and $P > 0$ such that

$$EP + PE^T = -R \tag{2}$$

for some given $R > 0$ ($R = I$ can always be chosen), where $E = A - BKC$. For the "if" direction, note that (2) implies the negativity of the real part of any eigenvalue of E, implying that the closed-loop E is stable. For the "only-if" direction, under the assumption that $E = A - BKC$ is stable for some given K, one can show that $P := \int_0^\infty \exp(Et)R\exp(E^T t)dt$ is well defined, satisfies $P^T = P$ and $P > 0$, and is the unique solution for (2).

Note that the set of all SOFs is given by $K = B^+XC^+ + L_BS + TR_C$ where S, T are any $m \times r$ matrices and X is any $n \times n$ matrix such that $E = A - BB^+XC^+C$ is stable. Thus, one can optimize K by utilizing the freeness in S, T without changing the closed-loop performance achieved by X. This characterization of the feasibility space shows its effectiveness in proving theorems, as will be seen along the chapter (see also [20, 35]). We also conclude that (A, B, C) is stabilizable if and only if (A, BB^+, C^+C) is stabilizable.

In the sequel, we make use of the following lemma (see [39]):

Lemma 2.1 The matrix equation $AX = B$ has solutions if and only if $AA^+B = B$ (equivalently, $R_AB = 0$). When the condition is satisfied, the set of all solutions is given by

$$X = A^+B + L_AZ, \tag{3}$$

where Z is arbitrary matrix. Moreover, we have: $\|X\|_F^2 = \|A^+B\|_F^2 + \|L_AZ\|_F^2$, implying that the minimal Frobenius-norm solution is $X = A^+B$.

Similarly, the equation $YA = B$ has solutions if and only if $BA^+A = B$ (equivalently, $BL_A = 0$). When the condition is satisfied, the set of all solutions is given by

$$Y = BA^+ + WR_A, \tag{4}$$

where W is arbitrary matrix. Moreover, we have: $\|Y\|_F^2 = \|BA^+\|_F^2 + \|WR_A\|_F^2$, implying that the minimal Frobenius-norm solution is $Y = BA^+$.

3. Parametrization of all the static state feedbacks

We start with the following lemma known as the projection lemma (see [41], Theorem 3.1):

Lemma 3.1 The pair (A, BB^+) is stabilizable if and only if there exists $P > 0$ such that

$$R_B(I + AP + PA^T)R_B = 0. \tag{5}$$

When (5) is satisfied then, X is a stabilizing SF if and only if X is a solution for

$$BB^+XP + PX^TBB^+ = I + AP + PA^T. \tag{6}$$

Moreover, one specific solution for (6) is given by

$$X_0 = \left(I + AP + PA^T\right)\left(I - \frac{1}{2}BB^+\right)P^{-1}. \tag{7}$$

Similarly, $\left(A^T, C^+C\right)$ is stabilizable if and only if there exists $Q > 0$ such that

$$L_C(I + A^TQ + QA)L_C = 0. \tag{8}$$

When (8) is satisfied then, Y^T is a stabilizing SF (i.e., $A^T - C^+CY^T$ or, equivalently, $A - YC^+C$ is stable) if and only if Y is a solution for:

$$C^+CY^TQ + QYC^+C = I + A^TQ + QA. \tag{9}$$

One specific solution for (9) is given by

$$Y_0 = Q^{-1}\left(I - \frac{1}{2}C^+C\right)\left(I + A^TQ + QA\right). \tag{10}$$

Remark 3.1 The explicit formulas (7) and (10) are our little contribution to the projection lemma, Lemma 3.1. Unfortunately, we do not have such an explicit formulas for LTI discrete-time systems.

In order to describe the set of all solutions for (6) and (9), we need the following lemma that can be proved easily:

Lemma 3.2 Let $P > 0, Q > 0$. Then, the set of all solutions for:

$$ZP + PZ^T = 0, \tag{11}$$

is given by $Z = WP^{-1}$ where $W^T = -W$.

Similarly, the set of all solutions for:

$$Z^T Q + QZ = 0, \tag{12}$$

is given by $Z = Q^{-1}V$ where $V^T = -V$.

The following theorem describes the set of all solutions for (6) and (9), using the controllers (7) and (10):

Theorem 3.1 Let $P > 0$ satisfy (5) and let X_0 be given by (7). Then, X is a solution for (6) if and only if:

$$X = X_0 + WP^{-1} + R_B L, \tag{13}$$

where W satisfies $W^T = -W, R_B W = 0$ and L is arbitrary.

Similarly, let $Q > 0$ satisfy (8) and let Y_0 be given by (10). Then, Y is a solution for (9) if and only if:

$$Y = Y_0 + Q^{-1}V + ML_C, \tag{14}$$

where V satisfies $V^T = -V, VL_C = 0$ and M is arbitrary.

Proof:

Assume that X is a solution for (6). Since X_0 is also a solution for (6), it follows that $BB^+(X - X_0)P + P(X - X_0)^T BB^+ = 0$. Let $Z = BB^+(X - X_0)$. Then, $R_B Z = 0$ and $ZP + PZ^T = 0$. Lemma 3.2 implies that $Z = WP^{-1}$, where $W^T = -W$ and therefore $R_B W = 0$. We conclude that $X - X_0 = WP^{-1} + R_B L$ for some L (namely, $L = X - X_0$).

Conversely, let X be given by (13) and let $Z = WP^{-1}$. Then, $BB^+(X - X_0) = BB^+ Z = Z$ since $BB^+ R_B = 0$ and since $R_B Z = 0$. Now, $ZP + PZ^T = 0$ implies that

$$BB^+(X - X_0)P + P(X - X_0)^T BB^+ = 0,$$

from which we conclude that X satisfies (6), since X_0 satisfies (6). The second claim is proved similarly. ■

In the following we describe the set $\mathcal{P}$ of all matrices $P > 0$ satisfying (5). Note that in Theorem 3.1 the existence of $P > 0$ satisfying (5) is guaranteed by the assumption that (A, BB^+) is stabilizable and as a result of Lemma 3.1. Let $P \in \mathcal{P}$ and let

$$\begin{cases} X_0 = (I + AP + PA^T) \cdot \left(I - \frac{1}{2}BB^+\right)P^{-1} \\ W \text{ arbitrary such that } W^T = -W, R_B W = 0 \\ X = X_0 + WP^{-1} + R_B L \text{ where } L \text{ is arbitrary} \\ K = B^+ X + L_B F \text{ where } F \text{ is arbitrary.} \end{cases} \tag{15}$$

Let $\mathcal{X}(P)$ denote the set of all matrices X satisfying (15) for a fixed $P \in \mathcal{P}$, and let $\mathcal{K}(P)$ denote the set of all matrices K satisfying (15) for a fixed $P \in \mathcal{P}$. Note that for a fixed $P \in \mathcal{P}$, the set $\mathcal{X}(P)$ is convex (actually affine) and $\cup_{P \in \mathcal{P}} \mathcal{X}(P)$ contains all the stabilizing X parameters of the stabilizable pair (A, BB^+). Finally, $\cup_{P \in \mathcal{P}} \mathcal{K}(P)$ contains all the stabilizing SF's K of the stabilizable pair (A, B).

For a stabilizable pair (A^T, C^T), let $\mathcal{Q}$ be the set of all matrices $Q > 0$ satisfying (8), and let

$$\begin{cases} Y_0 = Q^{-1} \cdot \left(I - \frac{1}{2} C^+ C\right)(I + A^T Q + QA) \\ V \text{ arbitrary such that } V^T = -V, VL_C = 0 \\ Y = Y_0 + Q^{-1} V + ML_C \text{ where } M \text{ is arbitrary} \\ K = YC^+ + GR_C \text{ where } G \text{ is arbitrary.} \end{cases} \tag{16}$$

Let $\mathcal{Y}(Q)$ denote the set of all matrices Y satisfying (16) for a fixed $Q \in \mathcal{Q}$, and let $\mathcal{K}(Q)$ denote the set of all matrices K satisfying (16) for a fixed $Q \in \mathcal{Q}$. Then, $\cup_{Q \in \mathcal{Q}} \mathcal{K}(Q)$ contains all the stabilizing SFs K of the stabilizable pair (A^T, C^T).

In the following we assume (without loss of generality, see Remark 4.2) that (A, BB^+) is controllable. Under this assumption, we recursively (go downwards and) define a sequence of sub-systems of the given system (A, BB^+). Since BB^+ is symmetric matrix (with simple eigenvalues from the set $\{0, 1\}$), it is diagonalizable by an orthogonal matrix. Let U denote an orthogonal matrix such that

$$\widehat{B} = U^T BB^+ U = \begin{bmatrix} I_k & 0 \\ 0 & 0 \end{bmatrix} = bdiag(I_k, 0) \tag{17}$$

(where $k = rank(B) = rank(BB^+) \geq 1$ since (A, B) is controllable). Let $\widehat{A} = U^T AU = \begin{bmatrix} \widehat{A}_{1,1} & \widehat{A}_{1,2} \\ \widehat{A}_{2,1} & \widehat{A}_{2,2} \end{bmatrix}$ be partitioned accordingly. Let $U^{(0)} = U$ and let $A^{(0)} = A, B^{(0)} = B, n_0 = n, k_0 = rank(B^{(0)})$. Similarly, let $U^{(1)}$ be an orthogonal matrix such that $U^{(1)T} B^{(1)} B^{(1)+} U^{(1)} = bdiag(I_{k_1}, 0)$, where $B^{(1)} = \widehat{A}_{2,1}$. Let $A^{(1)} = \widehat{A}_{2,2}, n_1 = n_0 - k_0, k_1 = rank(B^{(1)})$. Then, $\left(A^{(1)}, B^{(1)}\right)$ is controllable since $\left(A^{(0)}, B^{(0)}\right)$ is controllable (see [35] and see Lemma 5.1 in the following).

Recursively, assume that the pair $\left(A^{(i)}, B^{(i)}\right)$ was defined and is controllable. Let $U^{(i)}$ be an orthogonal matrix such that $\widehat{B^{(i)}} = U^{(i)T} B^{(i)} B^{(i)+} U^{(i)} = bdiag(I_{k_i}, 0)$, where $k_i \geq 1$ (since $\left(A^{(i)}, B^{(i)}\right)$ is controllable). Let $\widehat{A^{(i)}} = U^{(i)T} A^{(i)} U^{(i)} = \begin{bmatrix} \widehat{A^{(i)}}_{1,1} & \widehat{A^{(i)}}_{1,2} \\ \widehat{A^{(i)}}_{2,1} & \widehat{A^{(i)}}_{2,2} \end{bmatrix}$ be partitioned accordingly, with sizes $k_i \times k_i$ and $(n_i - k_i) \times (n_i - k_i)$ of the main diagonal blocks. Let $A^{(i+1)} = \widehat{A^{(i)}}_{2,2}, B^{(i+1)} = \widehat{A^{(i)}}_{2,1}, n_{i+1} = n_i - k_i, k_i = rank(B^{(i)})$. Then, $\left(A^{(i+1)}, B^{(i+1)}\right)$ is controllable. We stop the recursion when $B^{(i)} B^{(i)+} = I_{k_i}$ for some $i = b$ (i.e. the base case, in which also $k_b = n_b$).

Now, we go upward and define the Lyapunov matrices and the related SFs of the sub-systems. For the base case $i = b$, let $P^{(b)} > 0$ be arbitrary (note that it is a free parameter!). Let

$$\begin{cases} X_0^{(b)} = \frac{1}{2}\left(I_{n_b} + A^{(b)}P^{(b)} + P^{(b)}A^{(b)T}\right)\left(P^{(b)}\right)^{-1} \\ W^{(b)} \text{ arbitrary such that } W^{(b)T} = -W^{(b)} \\ X^{(b)} = X_0^{(b)} + W^{(b)}\left(P^{(b)}\right)^{-1} \\ K^{(b)} = B^{(b)+}X^{(b)} + L_{B^{(b)}}F^{(b)} \text{ where } F^{(b)} \text{ is arbitrary,} \end{cases} \tag{18}$$

and note that $R_{B^{(b)}} = 0$ in the base case. Now, it can be checked that $E^{(b)} = A^{(b)} - B^{(b)}K^{(b)} = A^{(b)} - X^{(b)}$ is stable. We therefore have a parametrization of $\mathcal{K}^{(b)}\left(P^{(b)}\right)$ through arbitrary $P^{(b)} > 0$.

Let $\mathcal{P}^{(i+1)}$ denote the set of all $P^{(i+1)} > 0$ satisfying:

$$R_{B^{(i+1)}}\left(I_{n_{i+1}} + A^{(i+1)}P^{(i+1)} + P^{(i+1)}A^{(i+1)T}\right)R_{B^{(i+1)}} = 0, \tag{19}$$

and assume that $\mathcal{K}^{(i+1)}\left(P^{(i+1)}\right)$ was parameterized through $P^{(i+1)} > 0$ ranging in the set $\mathcal{P}^{(i+1)}$, as is defined by (19). Similarly, let $\mathcal{P}^{(i)}$ denote the set of all $P^{(i)} > 0$ satisfying:

$$R_{B^{(i)}}\left(I_{n_i} + A^{(i)}P^{(i)} + P^{(i)}A^{(i)T}\right)R_{B^{(i)}} = 0, \tag{20}$$

and assume that $\mathcal{K}^{(i)}\left(P^{(i)}\right)$ was parameterized through $P^{(i)} > 0$ ranging in the set $\mathcal{P}^{(i)}$, as is defined by (20).

Now, we need to characterize the matrices $P^{(i)} > 0$ belonging to the set $\mathcal{P}^{(i)}$. Multiplying (20) from the left by $U^{(i)T}$ and from the right by $U^{(i)}$ we get:

$$\widehat{R_{B^{(i)}}}\left(I_{n_i} + \widehat{A^{(i)}}\widehat{P^{(i)}} + \widehat{P^{(i)}}\widehat{A^{(i)T}}\right)\widehat{R_{B^{(i)}}} = 0,$$

where $\widehat{R_{B^{(i)}}} = \begin{bmatrix} 0 & 0 \\ 0 & I_{n_i-k_i} \end{bmatrix}$ and $\widehat{P^{(i)}} = U^{(i)T}P^{(i)}U^{(i)} = \begin{bmatrix} \widehat{P^{(i)}}_{1,1} & \widehat{P^{(i)}}_{1,2} \\ \widehat{P^{(i)T}}_{1,2} & \widehat{P^{(i)}}_{2,2} \end{bmatrix}$ is partitioned accordingly. The condition (20) is therefore equivalent to:

$$\begin{aligned} & I_{n_i-k_i} + \left(\widehat{A^{(i)}}_{2,2} + \widehat{A^{(i)}}_{2,1}\widehat{P^{(i)}}_{1,2}\left(\widehat{P^{(i)}}_{2,2}\right)^{-1}\right)\widehat{P^{(i)}}_{2,2} + \\ & + \widehat{P^{(i)}}_{2,2}\left(\widehat{A^{(i)}}_{2,2} + \widehat{A^{(i)}}_{2,1}\widehat{P^{(i)}}_{1,2}\left(\widehat{P^{(i)}}_{2,2}\right)^{-1}\right)^T = 0, \end{aligned} \tag{21}$$

which is equivalent to:

$$\begin{aligned} & I_{n_i-k_i} + \left(A^{(i+1)} + B^{(i+1)}\widehat{P^{(i)}}_{1,2}\left(\widehat{P^{(i)}}_{2,2}\right)^{-1}\right)\widehat{P^{(i)}}_{2,2} + \\ & + \widehat{P^{(i)}}_{2,2}\left(A^{(i+1)} + B^{(i+1)}\widehat{P^{(i)}}_{1,2}\left(\widehat{P^{(i)}}_{2,2}\right)^{-1}\right)^T = 0. \end{aligned} \tag{22}$$

Let $P^{(i+1)} \in \mathcal{P}^{(i+1)}$ and let $K^{(i+1)} \in \mathcal{K}^{(i+1)}\left(P^{(i+1)}\right)$. Set $\widehat{P^{(i)}}_{2,2} := P^{(i+1)}$ and set $\widehat{P^{(i)}}_{1,2} := -K^{(i+1)}P^{(i+1)}$. Now, since $\widehat{P^{(i)}}_{2,2} := P^{(i+1)} > 0$, (22) implies that the system:

$$A^{(i+1)} + B^{(i+1)}\widehat{P^{(i)}}_{1,2}\left(\widehat{P^{(i)}}_{2,2}\right)^{-1} := A^{(i+1)} - B^{(i+1)}K^{(i+1)}, \tag{23}$$

is stable. Now,

$$\widehat{P^{(i)}} = \begin{bmatrix} \widehat{P^{(i)}}_{1,1} & -K^{(i+1)}P^{(i+1)} \\ -P^{(i+1)}K^{(i+1)T} & P^{(i+1)} \end{bmatrix} \tag{24}$$

and we need to define $\widehat{P^{(i)}}_{1,1}$ in order to complete $\widehat{P^{(i)}}$ to a strictly non-negative matrix. Since:

$$\begin{bmatrix} \widehat{P^{(i)}}_{1,1} & -K^{(i+1)}P^{(i+1)} \\ -P^{(i+1)}K^{(i+1)T} & P^{(i+1)} \end{bmatrix} =$$
$$= \begin{bmatrix} I_{k_i} & -K^{(i+1)} \\ 0 & I_{n_i-k_i} \end{bmatrix} \cdot$$
$$\cdot \begin{bmatrix} \widehat{P^{(i)}}_{1,1} - K^{(i+1)}P^{(i+1)}K^{(i+1)T} & 0 \\ 0 & P^{(i+1)} \end{bmatrix} \cdot$$
$$\cdot \begin{bmatrix} I_{k_i} & 0 \\ -K^{(i+1)T} & I_{n_i-k_i} \end{bmatrix},$$

it follows that $\widehat{P^{(i)}} > 0$ if and only if $\widehat{P^{(i)}}_{1,1} - K^{(i+1)}P^{(i+1)}K^{(i+1)T} > 0$ or equivalently if and only if $\widehat{P^{(i)}}_{1,1} = \Delta\widehat{P^{(i)}}_{1,1} + K^{(i+1)}P^{(i+1)}K^{(i+1)T}$, where $\Delta\widehat{P^{(i)}}_{1,1}$ is arbitrary strictly non-negative matrix (a free parameter!).

Conversely, if $P^{(i)} > 0$ satisfies (20) then (23) is stable and thus:

$$K^{(i+1)} = -\widehat{P^{(i)}}_{1,2}\left(\widehat{P^{(i)}}_{2,2}\right)^{-1} \in \mathcal{K}^{(i+1)}\left(R^{(i+1)}\right),$$

for some $R^{(i+1)} \in \mathcal{P}^{(i+1)}$. But since $K^{(i+1)} \in \mathcal{K}^{(i+1)}\left(R^{(i+1)}\right)$ if and only if:

$$I_{n_i-k_i} + \left(A^{(i+1)} - B^{(i+1)}K^{(i+1)}\right)R^{(i+1)} + R^{(i+1)}\left(A^{(i+1)} - B^{(i+1)}K^{(i+1)}\right)^T = 0, \tag{25}$$

since the last equation has unique strictly non-negative solution and since $\widehat{P^{(i)}}_{2,2}$ satisfies this equation, it follows that $R^{(i+1)} = \widehat{P^{(i)}}_{2,2}$. Let $P^{(i+1)} = \widehat{P^{(i)}}_{2,2}$. Then, $K^{(i+1)} \in \mathcal{K}^{(i+1)}\left(P^{(i+1)}\right)$ and since $\widehat{P^{(i)}}_{1,2} = -K^{(i+1)}P^{(i+1)}$, it follows that $\widehat{P^{(i)}}$ has the form (24). Thus, $\widehat{P^{(i)}}_{1,1} = \Delta\widehat{P^{(i)}}_{1,1} + K^{(i+1)}P^{(i+1)}K^{(i+1)T}$ where $\Delta\widehat{P^{(i)}}_{1,1} > 0$ is arbitrary (a free parameter!) and

$$\widehat{P^{(i)}} = U^{(i)T}P^{(i)}U^{(i)} = \begin{bmatrix} \Delta\widehat{P^{(i)}}_{1,1} + K^{(i+1)}P^{(i+1)}K^{(i+1)T} & -K^{(i+1)}P^{(i+1)} \\ -P^{(i+1)}K^{(i+1)T} & P^{(i+1)} \end{bmatrix}. \tag{26}$$

Therefore, $\mathcal{P}^{(i)}$ is the set of all $P^{(i)} > 0$ such that $\widehat{P^{(i)}} = U^{(i)T}P^{(i)}U^{(i)}$ is given by (26). We thus have a parametrization of all $P^{(i)} > 0$ satisfying (20). Specifically, $\mathcal{P}^{(0)}$ is the set of all $P^{(0)} > 0$ satisfying (5).

Now, let $P^{(i)} \in \mathcal{P}^{(i)}$ and let

$$\begin{cases} X_0^{(i)} = \left(I_{n_i} + A^{(i)}P^{(i)} + P^{(i)}A^{(i)T}\right)\cdot \\ \qquad \cdot\left(I_{n_i} - \frac{1}{2}B^{(i)}B^{(i)+}\right)\left(P^{(i)}\right)^{-1} \\ W^{(i)} \text{ arbitrary such that } W^{(i)T} = -W^{(i)}, R_{B^{(i)}}W^{(i)} = 0 \\ X^{(i)} = X_0^{(i)} + W^{(i)}\left(P^{(i)}\right)^{-1} + R_{B^{(i)}}L^{(i)} \text{ where } L^{(i)} \text{ is arbitrary} \\ K^{(i)} = B^{(i)+}X^{(i)} + L_{B^{(i)}}F^{(i)} \text{ where } F^{(i)} \text{ is arbitrary.} \end{cases} \tag{27}$$

Then, it can be checked that $E^{(i)} = A^{(i)} - B^{(i)}K^{(i)} = A^{(i)} - B^{(i)}B^{(i)+}X^{(i)}$ is stable. We therefore have a parametrization of $\mathcal{K}^{(i)}\left(P^{(i)}\right)$ through $P^{(i)} \in \mathcal{P}^{(i)}$. We conclude the discussion above with the following:

Theorem 3.2 Let (A, B) be a controllable pair. Then, in the above notations, for $i = b-1, \dots, 0$, $P^{(i)} > 0$ satisfies (20) if and only if $\widehat{P^{(i)}} = U^{(i)T}P^{(i)}U^{(i)}$ has the structure (26) where $\Delta\widehat{P^{(i)}}_{1,1} > 0$ is arbitrary (free parameter), where $K^{(i+1)} \in \mathcal{K}^{(i+1)}\left(P^{(i+1)}\right)$, where $P^{(b)} > 0$ is arbitrary (free parameter) and $\mathcal{K}^{(b)}\left(P^{(b)}\right)$ is given by (18). Moreover, $\mathcal{K}^{(i)}\left(P^{(i)}\right)$ for $i = b-1, \dots, 0$ is given by (27).

Similarly to the discussion above, relating to $\left(A^T, C^+C\right)$ and defining sub-systems for $j = 0, \dots, c$, we have a parametrization of all $Q^{(j)} > 0$ satisfying (8) for the related sub-system and specifically, $\mathcal{Q}^{(0)}$ is the set of all $Q^{(0)} > 0$ satisfying (8). The parametrizations of all the stabilizing SF's of (A, BB^+) and $\left(A^T, C^+C\right)$ are given in the following:

Corollary 3.1 Let (A, BB^+) be a given controllable pair. Then, the set of all stabilizing SF's of (A, BB^+) is given by $X = X_0 + WP^{-1} + R_BL$ where

$$X_0 = \left(I + AP + PA^T\right)\left(I - \frac{1}{2}BB^+\right)P^{-1},$$

where L is arbitrary, W satisfies $W^T = -W, R_BW = 0$, and $P > 0$ satisfies

$$R_B\left(I + AP + PA^T\right)R_B = 0,$$

i.e. $P \in \mathcal{P}^{(0)}$.

Similarly, let $\left(A^T, C^+C\right)$ be a given controllable pair. Then, the set of all stabilizing SF's of $\left(A^T, C^+C\right)$ is given by $Y = Y_0 + Q^{-1}V + ML_C$ where

$$Y_0 = Q^{-1}\left(I - \frac{1}{2}C^+C\right)\left(I + A^TQ + QA\right),$$

where M is arbitrary, V satisfies $V^T = -V, VL_C = 0$, and $Q > 0$ satisfies

$$L_C\left(I + A^TQ + QA\right)L_C = 0,$$

i.e. $Q \in \mathcal{Q}^{(0)}$.

4. Parametrizations of all the static output feedbacks

In this section, we give two parametrizations for the set of all the stabilizing SOFs. We start with the following lemma, which was extensively used in [20]:

Lemma 4.1 A system (A, B, C) is stabilizable if and only if (A, B) and $\left(A^T, C^T\right)$ are stabilizable and there exist matrices $X, Y \in \mathbb{R}^{n \times n}$ such that $A - BB^+X$ and $A - YC^+C$ are stable and $BB^+X = YC^+C$. When the conditions hold, the set of all stabilizing SOFs related to the chosen matrices X, Y is given by $K_X = B^+XC^+ + L_BS + TR_C$ or by $K_Y = B^+YC^+ + L_BF + HR_C$ respectively, where S, T, F, H are any $m \times r$ matrices. The closed-loop matrix is given by $E = A - BK_XC = A - BB^+X = A - YC^+C = A - BK_YC$.

Remark 4.1 Under the hypotheses of Corollary 3.1, note that $X = X_0 + WP^{-1} + R_BL$ and $Y = Y_0 + Q^{-1}V + ML_C$ satisfies $BB^+X = YC^+C$ if and only if $BB^+X_0 + WP^{-1} = Y_0C^+C + Q^{-1}V$, since $BB^+W = W, BB^+R_B = 0, VC^+C = V, L_CC^+C = 0$. Moreover, this condition can be simplified to (meaning that it does not include matrix inverses):

$$\begin{aligned} &QBB^+\left(I + AP + PA^T\right)\left(I - \frac{1}{2}BB^+\right) + QW = \\ &= \left(I - \frac{1}{2}C^+C\right)\left(I + A^TQ + QA\right)C^+CP + VP. \end{aligned} \tag{28}$$

We can state now the first parametrization for the set of all the stabilizing SOFs:

Corollary 4.1 Let (A, B, C) be a given system triplet. Assume that (A, B), $\left(A^T, C^T\right)$ are controllable. Then, the system has a stabilizing static output feedback if and only if there exist $P, Q > 0$ and W, V such that

$$\begin{cases} R_B\left(I + AP + PA^T\right)R_B = 0 \ \left(i.e. P \in \mathcal{P}^{(0)}\right) \\ L_C\left(I + A^TQ + QA\right)L_C = 0 \ \left(i.e. Q \in \mathcal{Q}^{(0)}\right) \\ W^T = -W, R_BW = 0 \\ V^T = -V, VL_C = 0 \\ QBB^+\left(I + AP + PA^T\right)\left(I - \frac{1}{2}BB^+\right) + QW = \\ = \left(I - \frac{1}{2}C^+C\right)\left(I + A^TQ + QA\right)C^+CP + VP. \end{cases}$$

In this case, $A - BKC$ is stable if and only if

$$\begin{cases} K = K_X = B^+XC^+ + L_BS + TR_C \\ X = X_0 + WP^{-1} + R_BL \\ X_0 = \left(I + AP + PA^T\right)\left(I - \frac{1}{2}BB^+\right)P^{-1}, \end{cases}$$

where S, T, L are arbitrary.

Similarly, $A - BKC$ is stable if and only if

$$\begin{cases} K = K_Y = B^+YC^+ + L_BF + HR_C \\ Y = Y_0 + Q^{-1}V + ML_C \\ Y_0 = Q^{-1}\left(I - \frac{1}{2}C^+C\right)(I + A^TQ + QA), \end{cases}$$

where F, H, M are arbitrary.

We conclude this section with a second SOF parametrization:

Corollary 4.2 Let (A, B) and (A^T, C^T) be controllable pairs. Then, $A - BKC$ is stable if and only if there exists $K^{(0)} \in \mathcal{K}^{(0)}(P^{(0)})$ for some $P^{(0)} \in \mathcal{P}^{(0)}$, such that $K^{(0)}L_C = 0$. In this case, the set of all K's such that $A - BKC$ is stable, is given by $K = K^{(0)}C^+ + GR_C$ where $K^{(0)} \in \mathcal{K}^{(0)}(P^{(0)})$, and G is arbitrary.

Proof: If there exists $K^{(0)} \in \mathcal{K}^{(0)}(P^{(0)})$ such that $K^{(0)}L_C = 0$ for some $P^{(0)} \in \mathcal{P}^{(0)}$ then $K^{(0)} = K^{(0)}C^+C$. Since $K^{(0)} \in \mathcal{K}^{(0)}(P^{(0)})$ it follows that $A - BK^{(0)}$ is stable. Thus, for $K = K^{(0)}C^+$ we get that $A - BKC$ is stable.

Conversely, if $A - BKC$ is stable for some K then, for $K^{(0)} = KC$ we have $K^{(0)}L_C = 0$ and since $A - BK^{(0)}$ is stable, Theorem 3.2 implies that there exists $P^{(0)} \in \mathcal{P}^{(0)}$ such that $K^{(0)} \in \mathcal{K}^{(0)}(P^{(0)})$. ■.

Remark 4.2 Note that when (A, BB^+) is stabilizable, there exists an orthogonal matrix V such that

$$V^TAV = \begin{bmatrix} \widehat{A}_{1,1} & 0 \\ \widehat{A}_{2,1} & \widehat{A}_{2,2} \end{bmatrix}, V^TBB^+V = \begin{bmatrix} 0 & 0 \\ 0 & \widehat{B}_{2,2}\widehat{B}_{2,2}^+ \end{bmatrix},$$

where $\widehat{A}_{1,1}$ is stable and $\left(\widehat{A}_{2,2}, \widehat{B}_{2,2}\widehat{B}_{2,2}^+\right)$ is controllable (see [35], Lemma 3.1, p. 536). Thus, we may assume without loss of generality that the given pair is controllable. If (A, B, C) is a system triplet such that (A, BB^+) and (A^T, C^+C) are stabilizable then, there exists an orthogonal matrix V such that $\left(\widehat{A}_{2,2}, \widehat{B}_{2,2}\widehat{B}_{2,2}^+\right)$ and $\left(\widehat{A}_{2,2}^T, \widehat{C}_{2,2}^+\widehat{C}_{2,2}\right)$ are controllable, where $\widehat{C} = V^TC^+CV$ is partitioned accordingly (see [35], Theorem 4.1 and Remark 4.1, p. 539). Thus, the assumption in Corollary 4.2 that the given pairs are controllable does not make any loss of generality of the results.

The effectiveness of the method is shown in the following example, but first, for the convenience of the reader, we summarize the whole method in Algorithm 1 (with its continuation in Algorithm 2). Let $f(K)$ denote a target function of the SOF K, to be minimized (e.g., $\|K\|_F$, the LQR functional, the H_∞-norm or the H_2-norm of the closed loop, the pole-placement errors of the closed loop, or any other key performance that depends on K).

Regarding the LQR problem, let the LQR functional be defined by:

$$J(x_0, u) = \int_0^\infty \left(x(t)^TQx(t) + u(t)^TRu(t)\right)dt, \tag{29}$$

where $Q > 0$ and $R \geq 0$ are given. We need to find $u(t)$ that minimizes the functional value for any initial disturbance x_0 from the equilibrium point 0. Assuming that $u(t)$ is realized by a stabilizing SOF, let $u(t) = -Ky(t) = -KCx(t)$. Then, by substitution of the last into (29), we get:

$$J(x_0, K) = \int_0^\infty x(t)^T(Q + C^TK^TRKC)x(t)dt. \tag{30}$$

Now, since $Q + C^T K^T RKC > 0$ and since $E := A - BKC$ is stable, the Lyapunov equation:

$$E^T P + PE = -(Q + C^T K^T RKC), \tag{31}$$

has unique solution $P_{LQR}(K) > 0$ given by:

$$\begin{aligned} P_{LQR}(K) &= \int_0^\infty \exp(E^T t)(Q + C^T K^T RKC)\exp(Et)dt = \\ &= -mat\left(\left(I \otimes E^T + E^T \otimes I\right)^{-1}\right)vec(Q + C^T K^T RKC). \end{aligned} \tag{32}$$

By substitution of (31) into (30), we get:

$$J(x_0, K) = x_0^T P_{LQR}(K) x_0 = \left\| P_{LQR}(K)^{\frac{1}{2}} x_0 \right\|_2^2. \tag{33}$$

Algorithm 1. An Algorithm For Optimal SOF's.

Require: *An algorithm for optimizing $f(K)$ under LMI and linear constraints, an algorithm for computing the Moore-Penrose pseudo-inverse and an algorithm for orthogonal diagonalization.*

Input: *System triplet (A, B, C) such that (A, B), (A^T, C^T) are controllable.*

Output: *SOF K such that $A - BKC$ is stable minimizing $f(K)$— if exists*

1. $A^{(0)} \leftarrow A$
2. $B^{(0)} \leftarrow B$
3. $i \leftarrow 0$
4. $k_0 \leftarrow rank\left(B^{(0)}\right)$
5. **while** $B^{(i)} B^{(i)+} \neq I_{k_i}$ **do**
6. *compute orthogonal matrix* $U^{(i)}$ *such that* $U^{(i)T} B^{(i)} B^{(i)+} U^{(i)} = bdiag(I_{k_i}, 0)$
7. $\widehat{A^{(i)}} \leftarrow U^{(i)T} A^{(i)} U^{(i)}$
8. *partition* $\widehat{A^{(i)}} = \begin{bmatrix} \widehat{A^{(i)}}_{1,1} & \widehat{A^{(i)}}_{1,2} \\ \widehat{A^{(i)}}_{2,1} & \widehat{A^{(i)}}_{2,2} \end{bmatrix}$
9. $A^{(i+1)} \leftarrow \widehat{A^{(i)}}_{2,2}$
10. $B^{(i+1)} \leftarrow \widehat{A^{(i)}}_{2,1}$
11. $i \leftarrow i + 1$
12. $k_i \leftarrow rank\left(B^{(i)}\right)$
13. **end while**
14. $b \leftarrow i$
15. *let* $P^{(b)}$ *be a symbol for* $P^{(b)} > 0$
16. $X_0^{(b)} \leftarrow \frac{1}{2}\left(I_{n_b} + A^{(b)} P^{(b)} + P^{(b)} A^{(b)T}\right)\left(P^{(b)}\right)^{-1}$
17. *let* $W^{(b)}$ *be a symbol for a matrix satisfying* $W^{(b)T} = -W^{(b)}$
18. $X^{(b)} \leftarrow X_0^{(b)} + W^{(b)}\left(P^{(b)}\right)^{-1}$
19. *let* $F^{(b)}$ *be a symbol for arbitrary matrix*
20. $K^{(b)} \leftarrow B^{(b)+} X^{(b)} + L_{B^{(b)}} F^{(b)}$

Thus,

$$
\begin{aligned}
J(x_0, K) &= \left\| P_{LQR}(K)^{\frac{1}{2}} x_0 \right\|_2^2 \leq \\
&\leq \left\| P_{LQR}(K)^{\frac{1}{2}} \right\|^2 \|x_0\|_2^2 = \\
&= \|P_{LQR}(K)\| \|x_0\|_2^2 = \\
&= \sigma_{max}(P_{LQR}(K)) \|x_0\|_2^2,
\end{aligned}
$$

Algorithm 2. An Algorithm For Optimal SOF's, Continued.

1. **for** $i = b - 1$ *downto* 0 **do**
2. *let* $\widehat{\Delta P^{(i)}}_{1,1}$ *be a symbol for* $\widehat{\Delta P^{(i)}}_{1,1} > 0$
3. $\widehat{P^{(i)}} \leftarrow \begin{bmatrix} \widehat{\Delta P^{(i)}}_{1,1} + K^{(i+1)} P^{(i+1)} K^{(i+1)T} & -K^{(i+1)} P^{(i+1)} \\ -P^{(i+1)} K^{(i+1)T} & P^{(i+1)} \end{bmatrix}$
4. $\left(\widehat{P^{(i)}}\right)^{-1} \leftarrow \begin{bmatrix} \left(\widehat{\Delta P^{(i)}}_{1,1}\right)^{-1} & \left(\widehat{\Delta P^{(i)}}_{1,1}\right)^{-1} K^{(i+1)} \\ K^{(i+1)T} \left(\widehat{\Delta P^{(i)}}_{1,1}\right)^{-1} & \left(P^{(i+1)}\right)^{-1} + K^{(i+1)T} \left(\widehat{\Delta P^{(i)}}_{1,1}\right)^{-1} K^{(i+1)} \end{bmatrix}$
5. $P^{(i)} \leftarrow U^{(i)} \widehat{P^{(i)}} U^{(i)T}$
6. $\left(P^{(i)}\right)^{-1} \leftarrow U^{(i)} \left(\widehat{P^{(i)}}\right)^{-1} U^{(i)T}$
7. $X_0^{(i)} \leftarrow \left(I_{n_i} + A^{(i)} P^{(i)} + P^{(i)} A^{(i)T}\right) \left(I_{n_i} - \frac{1}{2} B^{(i)} B^{(i)+}\right) \left(P^{(i)}\right)^{-1}$
8. *let* $W^{(i)}$ *be a symbol for a matrix satisfying* $W^{(i)T} = -W^{(i)}$ *and* $R_{B^{(i)}} W^{(i)} = 0$
9. *let* $L^{(i)}$ *be a symbol for arbitrary matrix*
10. *let* $F^{(i)}$ *be a symbol for arbitrary matrix*
11. $X^{(i)} \leftarrow X_0^{(i)} + W^{(i)} \left(P^{(i)}\right)^{-1} + R_{B^{(i)}} L^{(i)}$
12. $K^{(i)} \leftarrow B^{(i)+} X^{(i)} + L_{B^{(i)}} F^{(i)}$
13. **end for**
14. *optimize* $f(K)$ *under the matrix equation* $K^{(0)} L_C = 0$
    and the constraints $\widehat{\Delta P^{(0)}}_{1,1} > 0, \ldots, \widehat{\Delta P^{(b-1)}}_{1,1} > 0, P^{(b)} > 0$
    with respect to $F^{(0)}, \ldots, F^{(b)}$ *to* $W^{(0)}, \ldots, W^{(b)}$
    and to $\widehat{\Delta P^{(0)}}_{1,1}, \ldots, \widehat{\Delta P^{(b-1)}}_{1,1}, P^{(b)}$ *as variables*
15. **if** *a solution was found* **then**
16. **return** K
17. **else**
18. **return** *"no solution was found"*
19. **end if**

where $\sigma_{max}(P_{LQR}(K))$ is the largest eigenvalue of $P_{LQR}(K)$. Therefore,

$$
\frac{J(x_0, K)}{\|x_0\|_2^2} \leq \sigma_{max}(P_{LQR}(K)). \tag{34}
$$

Now, if x_0 is known then we can minimize $J(x_0, K)$ by minimizing $x_0^T P_{LQR}(K) x_0$. Otherwise, and if we design for the worst-case, we need to minimize $\sigma_{max}(P_{LQR}(K))$.

In the following examples, we have executed the algorithm on Processor: Intel (R) Core(TM) i5-2400 CPU @ 3.10GHz 3.10 GHz, RAM: 8.00 GB, Operating System: Windows 10, System Type: 64-bit Operating System, x64-based processor, Platform: MATLAB®, Version: R2018b, Function: fmincon.

Example 4.1 A system of Boeing B-747 aircraft (the "AC5" system in [42] and see also [43]) is given by the general model (given here with slight changes):

$$\begin{cases} \frac{d}{dt}x(t) = Ax(t) + B_1w(t) + Bu(t) \\ z(t) = C_1x(t) + D_{1,1}w(t) + D_{1,2}u(t) \\ y(t) = Cx(t) + D_{2,1}w(t), \end{cases}$$

where x is the state, w is the noise, u is the control input, z is the regulated output, and y is the measurement, where:

$$A = \begin{bmatrix} 0.980100000000000 & 0.000300000000000 & -0.098000000000000 & 0.003800000000000 \\ -0.386800000000000 & 0.907100000000000 & 0.047100000000000 & -0.000800000000000 \\ 0.159100000000000 & -0.001500000000000 & 0.969100000000000 & 0.000300000000000 \\ -0.019800000000000 & 0.095800000000000 & 0.002100000000000 & 1.000000000000000 \end{bmatrix}$$

$$B = \begin{bmatrix} -0.000100000000000 & 0.005800000000000 \\ 0.029600000000000 & 0.015300000000000 \\ 0.001200000000000 & -0.090800000000000 \\ 0.001500000000000 & 0.000800000000000 \end{bmatrix}, C = \begin{bmatrix} 1 & 0 & 0 & 0 \\ 0 & 0 & 0 & 1 \end{bmatrix},$$

$B_1 = B, C_1 = C, D_{1,1} = 0_4, D_{2,1} = 0_{2\times 4}, D_{1,2} = 0_{4\times 2}$.

with

$$\sigma(A) = \begin{Bmatrix} 0.978871342065923 \pm 0.128159143146289i \\ 0.899614120838404 \\ 0.998943195029751 \end{Bmatrix}.$$

Note that (A, B) and $\left(A^T, C^T\right)$ here are controllable. Let $u = u_r - Ky$, where u_r is a reference input. Then, $u = u_r - KCx - KD_{2,1}w$ and substitution the last into the system yields the closed-loop system:

$$\begin{cases} \frac{d}{dt}x(t) = (A - BKC)x(t) + (B_1 - BKD_{2,1})w(t) + Bu_r(t) \\ z(t) = (C_1 - D_{1,2}KC)x(t) + (D_{1,1} - D_{1,2}KD_{2,1})w(t) + D_{1,2}u_r(t), \end{cases}$$

where the behavior of z is of our interest. Note that we actually have:

$$\begin{cases} \frac{d}{dt}x(t) = (A - BKC)x(t) + Bw(t) + Bu_r(t) \\ z(t) = Cx(t) = y(t). \end{cases}$$

For the stabilization via SOF with minimal Frobenius-norm, we need to minimize $f(K) = \|K\|_F$. For the LQR problem we need to minimize $f(K) = x_0^T P_{LQR}(K)x_0$ when x_0 is known and to minimize $f(K) = \sigma_{max}(P_{LQR}(K))$ when x_0 is unknown, where $P_{LQR}(K)$ is given by (32). For the H_∞ and the H_2 problems, we need to minimize $f(K) = \|T_{w,z}(s)\|_{H_\infty}$ and $f(K) = \|T_{w,z}(s)\|_{H_2}$, resp. where:

$$T_{w,z}(s) = (D_{1,1} - D_{1,2}KD_{2,1}) + (C_1 - D_{1,2}KC)(sI - A + BKC)^{-1}(B_1 - BKD_{2,1}) = \\ = C(sI - A + BKC)^{-1}B.$$

These problems needed to be solved under the constraint that $A - BKC$ stable, i.e., that $K = K^{(0)}C^{+} + GR_C$, where $K^{(0)} \in K^{(0)}\left(P^{(0)}\right)$ for some $P^{(0)} \in P^{(0)}$, such that $K^{(0)}L_C = 0$.

Applying the algorithm we had:

$$U^{(0)} = \begin{bmatrix} -0.063882699439918 & 0 & -0.997957414277919 & 0 \\ 0.012195875662698 & 0.998643130545343 & -0.000780700106257 & -0.050621625208610 \\ 0.997882828566422 & -0.012222870716218 & -0.063877924951025 & 0.000269421014890 \\ 0.000348971870720 & 0.050621134381318 & -0.000022338894237 & 0.998717867304654 \end{bmatrix}$$

$$A^{(1)} = \begin{bmatrix} 0.983650838535766 & -0.003772277911855 \\ 0.000091490905229 & 0.994959072276129 \end{bmatrix}, B^{(1)} = \begin{bmatrix} 0.098883972659475 & -0.001544978964136 \\ 0.000939225045070 & 0.100248978115558 \end{bmatrix}.$$

The "while-loop" stops because $B^{(1)}B^{(1)+} = I_2$. We have

$$B^{(0)+} = \begin{bmatrix} -0.386571795892900 & 33.468356299440529 & 5.629731696802776 & 1.694878880538571 \\ 0.696955535886478 & 0.442712098375399 & -10.893875111014371 & 0.025378383262705 \end{bmatrix}$$

$$B^{(1)+} = \begin{bmatrix} 10.111382188357680 & 0.155830743345230 \\ -0.094732770050116 & 9.973704058215121 \end{bmatrix}$$

$L_{B^{(0)}} = 0_2, R_{C^{(0)}} = 0_2, L_{B^{(1)}} = 0_2,$

$$R_{B^{(0)}} = \begin{bmatrix} 0.995919000712269 & 0.000779105459367 & 0.063747448813564 & 0.000022293265130 \\ 0.000779105459367 & 0.002563158431417 & 0.000036230973158 & -0.050556704127861 \\ 0.063747448813564 & 0.000036230973158 & 0.004080461883732 & 0.000270502543607 \\ 0.000022293265130 & -0.050556704127861 & 0.000270502543607 & 0.997437378972582 \end{bmatrix}, R_{B^{(1)}} = 0_2$$

$$I_4 - \frac{1}{2}B^{(0)}B^{(0)+} = \begin{bmatrix} 0.997959500356135 & 0.000389552729683 & 0.031873724406782 & 0.000011146632565 \\ 0.000389552729683 & 0.501281579215708 & 0.000018115486579 & -0.025278352063931 \\ 0.031873724406782 & 0.000018115486579 & 0.502040230941866 & 0.000135251271804 \\ 0.000011146632565 & -0.025278352063931 & 0.000135251271804 & 0.998718689486291 \end{bmatrix}$$

$I_2 - \frac{1}{2}B^{(1)}B^{(1)+} = \frac{1}{2}I_2.$

Now, we parameterize all the matrices $K^{(0)}$ such that $A^{(0)} - B^{(0)}K^{(0)}$ is stable. Let $P^{(1)} = \begin{bmatrix} p_1 & p_2 \\ p_2 & p_3 \end{bmatrix}$, where $p_1, d_1 := p_1p_3 - p_2^2 > 0$. Let w_1 be arbitrary and let $W^{(1)} = \begin{bmatrix} 0 & w_1 \\ -w_1 & 0 \end{bmatrix}$. Let

$$S^{(1)} = B^{(1)+}\left(\frac{1}{2}\left(I_2 + A^{(1)}P^{(1)} + P^{(1)}A^{(1)T}\right) + W^{(1)}\right).$$

Then

$$K^{(1)} = S^{(1)}\left(P^{(1)}\right)^{-1}.$$

Let $\widehat{\Delta P^{(0)}}_{1,1} = \begin{bmatrix} p_4 & p_5 \\ p_5 & p_6 \end{bmatrix}$, where $p_4, d_2 := p_4 p_6 - p_5^2 > 0$. Then,

$$P^{(0)} = U^{(0)} \begin{bmatrix} \widehat{\Delta P^{(0)}}_{1,1} + K^{(1)} P^{(1)} K^{(1)T} & -K^{(1)} P^{(1)} \\ -P^{(1)} K^{(1)T} & P^{(1)} \end{bmatrix} U^{(0)T} =$$

$$= U^{(0)} \begin{bmatrix} \widehat{\Delta P^{(0)}}_{1,1} + S^{(1)} K^{(1)T} & -S^{(1)} \\ -S^{(1)T} & P^{(1)} \end{bmatrix} U^{(0)T},$$

and

$$\left(P^{(0)}\right)^{-1} = U^{(0)} \begin{bmatrix} \left(\widehat{\Delta P^{(0)}}_{1,1}\right)^{-1} & \left(\widehat{\Delta P^{(0)}}_{1,1}\right)^{-1} K^{(1)} \\ K^{(1)T} \left(\widehat{\Delta P^{(0)}}_{1,1}\right)^{-1} & \left(P^{(1)}\right)^{-1} + K^{(1)T} \left(\widehat{\Delta P^{(0)}}_{1,1}\right)^{-1} K^{(1)} \end{bmatrix} U^{(0)T}.$$

Let

$$S^{(0)} = B^{(0)+} \left(I_4 + A^{(0)} P^{(0)} + P^{(0)} A^{(0)T}\right) \left(I_4 - \frac{1}{2} B^{(0)} B^{(0)+}\right).$$

Then

$$K^{(0)} = S^{(0)} \left(P^{(0)}\right)^{-1}.$$

Note that $W^{(0)T} = -W^{(0)}, R_{B^{(0)}} W^{(0)} = 0$ implies that $W^{(0)} = 0_4$. We have completed the parametrization of all the SFs of the system, where the parameters $W^{(1)}$, $P^{(1)} > 0$, and $\widehat{\Delta P^{(0)}}_{1,1} > 0$ are free.

Regarding the optimization stage, we had the following results: starting from the point (feasible for SF but not feasible for SOF):

$$\begin{bmatrix} p_1 & p_2 & p_3 & p_4 & p_5 & p_6 & w_1 \end{bmatrix} =$$
$$= \begin{bmatrix} 750 & 0 & 750 & 750 & 0 & 750 & 0 \end{bmatrix},$$

in $CPU - Time = 0.59375[sec]$ the fmincon function (with the interior-point option and the default optimization parameters) has converged to the optimal point

$$\begin{bmatrix} p_1 & p_2 & p_3 & p_4 & p_5 & p_6 & w_1 \end{bmatrix} =$$
$$= 10^7 \cdot \begin{bmatrix} 0.009103654227197 & 0.000105735816664 & 0.006486122495094 & 1.912355216342858 & 0.057053301125719 & 1.792930229298237 & -0.000647092932030 \end{bmatrix},$$

resulting with the following optimal Frobenius-norm SF and SOF

$$K^{(0)} = 10^3 \cdot \begin{bmatrix} -0.157533999747776 & -0.000000000000000 & -0.000000000000000 & 1.273776653891793 \\ 0.332650954180600 & -0.000000000000000 & -0.000000000000000 & 0.000655866128882 \end{bmatrix}$$

$$K = 10^3 \cdot \begin{bmatrix} -0.157533999747776 & 1.273776653891793 \\ 0.332650954180600 & 0.000655866128882 \end{bmatrix},$$

with $\|K\|_F = 1.325888763265586 \cdot 10^3$. The resulting closed-loop eigenvalues are:

$$\sigma\left(A^{(0)} - B^{(0)}KC^{(0)}\right) = \left\{ \begin{matrix} -0.000004083911866 \pm 1.423412274467895i \\ -0.000005220069661 \pm 1.681989978268793i \end{matrix} \right\}.$$

For a comparison, in this (small) example we had seven scalar indeterminate, four scalar equations and four scalar inequalities while by the BMI method $\left(A^{(0)} - B^{(0)}KC^{(0)}\right)P + P\left(A^{(0)} - B^{(0)}KC^{(0)}\right)^T < 0, P > 0$ we would have 14 scalar indeterminate and eight scalar inequalities. This shows the potential of the method in reducing the number of variables and inequalities/equations, thus enabling to deal efficiently with larger problems. Moreover, the method removes the decoupling of P and K, in the sense that now K depends on P and the dependence of P on K has removed, thus making the problem more relaxed.

Figures 1 and **2** show the impulse response and the step response of the closed-loop system, in terms of the regulated output $z = y$, where $w = 0$ and u_r is the delta Dirac function or the unit-step function, respectively. While the amplitudes seem to be reasonable, the settling time of order 10^5 seems unreasonable. This happens because lowering the SOF-norm results in pushing the closed-loop eigenvalues toward the imaginary axes, as can be seen from the dense oscillations. We therefore must set a barrier on the abscissa of the closed-loop eigenvalues as a constraint. Note however that as a starting point for other optimization keys where we need any stabilizing SOF that we can get, the above SOF might be sufficient.

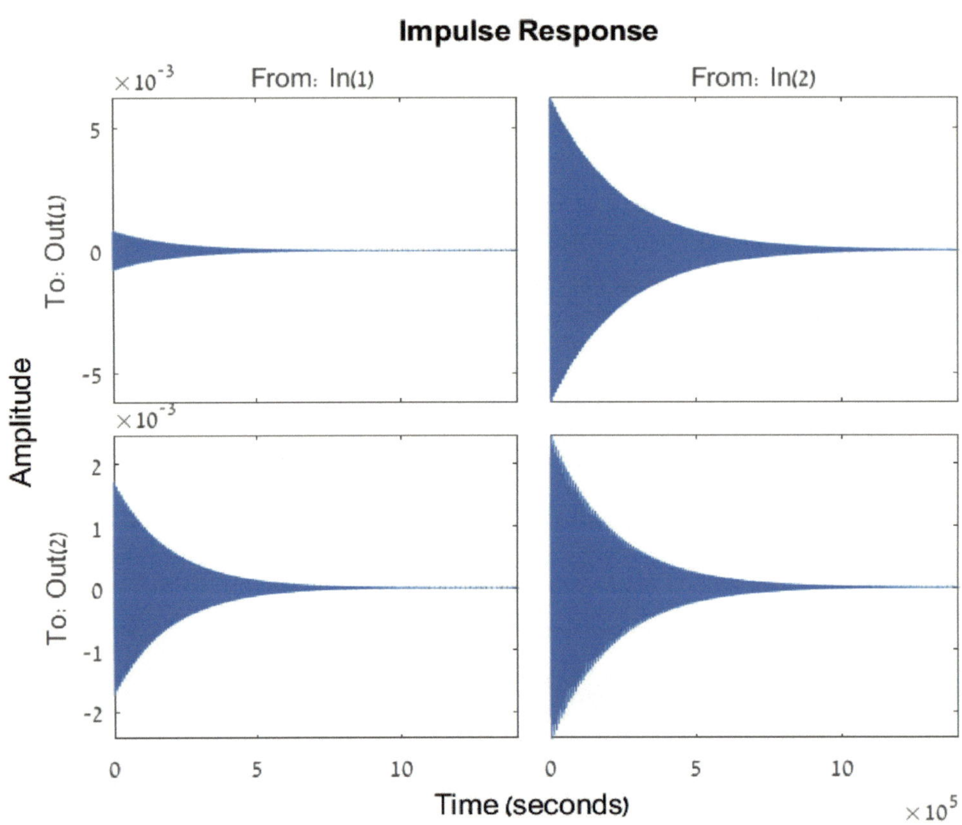

Figure 1.
Impulse response of the closed loop with the minimal-norm SOF.

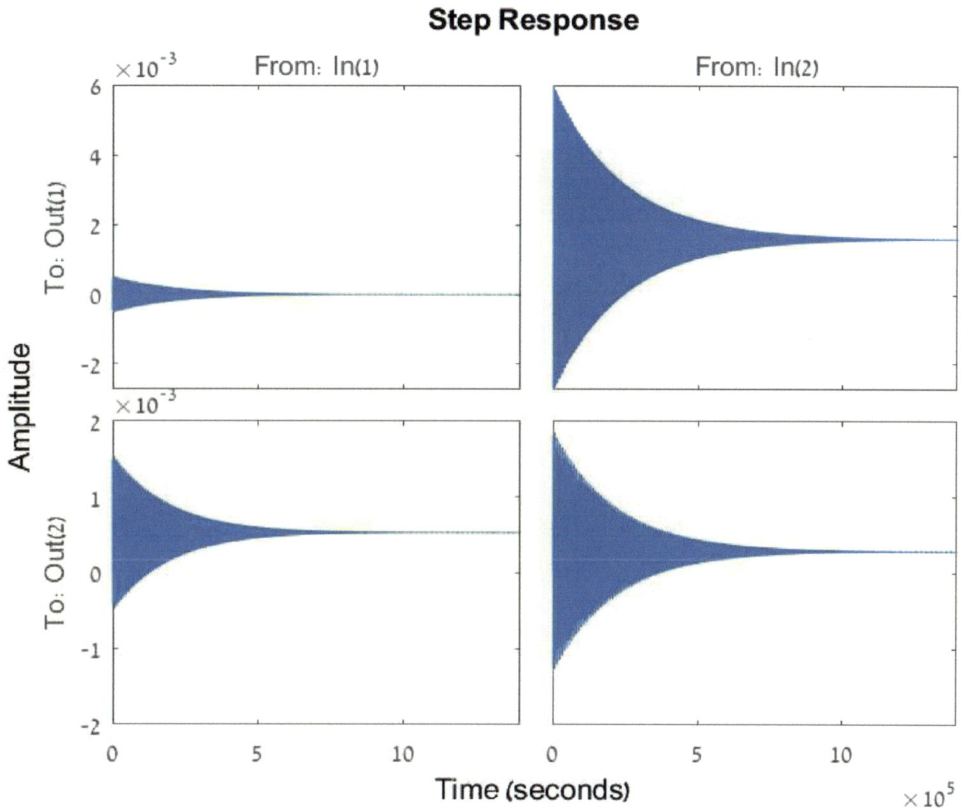

Figure 2.
Step response of the closed loop with the minimal-norm SOF.

Regarding the LQR functional with $Q = I, R = I$, starting from:

$$\begin{bmatrix} p_1 & p_2 & p_3 & p_4 & p_5 & p_6 & w_1 \end{bmatrix} = \\ = \begin{bmatrix} 300 & 0 & 300 & 300 & 0 & 300 & 15 \end{bmatrix},$$

in $CPU - Time = 0.90625[sec]$ the fmincon function has converged to the optimal point

$$\begin{bmatrix} p_1 & p_2 & p_3 & p_4 & p_5 & p_6 & w_1 \end{bmatrix} = \\ = 10^2 \cdot \begin{bmatrix} 0.010603084813420 & -0.002595549083521 & 0.009614240830002 & 1.009076872432389 & -0.832493933321482 & 1.812148701422345 & 0.001750170924431 \end{bmatrix},$$

resulting with the following optimal SF and SOF

$$K^{(0)} = 10^3 \cdot \begin{bmatrix} -1.466094499085196 & 0.000000000000002 & -0.000000000000000 & 2.731682559443639 \\ 0.703352598722306 & 0.000000000000000 & 0.000000000000001 & -0.149101634953946 \end{bmatrix}$$

$$K = 10^3 \cdot \begin{bmatrix} -1.466094499085196 & 2.731682559443639 \\ 0.703352598722306 & -0.149101634953946 \end{bmatrix},$$

with $\|K\|_F = 3.182523976577676 \cdot 10^3$ and LQR wort-case functional-value $\sigma_{max}(P_{LQR}(K)) = 1.981249586261248 \cdot 10^6$. The resulting closed-loop eigenvalues are:

$$\sigma\left(A^{(0)} - B^{(0)}KC^{(0)}\right) = \left\{ \begin{array}{c} -1.723892066022943 \pm 1.346849871126735i \\ -0.450106460827155 \pm 1.711908728695912i \end{array} \right\}.$$

The entries of $z = y$ under $w = 0$ and $u_r = 0$, when the closed-loop system is derived by the initial condition $x_0 = \begin{bmatrix} 1 & 1 & 1 & 1 \end{bmatrix}^T$ are depicted in **Figure 3**. The results might not be satisfactory regarding the amplitudes or the settling time; however, as a starting point for other optimization keys where we need any stabilizing SOF that we can get, the above SOF might be sufficient.

For the problem of pole placement via SOF, assume that the target is to place the closed-loop eigenvalue as close as possible to $-10 \pm i, \; -1 \pm 0.1i$. Then, starting from:

$$\begin{bmatrix} p_1 & p_2 & p_3 & p_4 & p_5 & p_6 & w_1 \end{bmatrix} = \\ = \begin{bmatrix} 1 & 0 & 1 & 1 & 0 & 1 & 0 \end{bmatrix},$$

in $CPU - Time = 1.828125[sec]$ the fmincon function has converged to the optimal point

$$\begin{bmatrix} p_1 & p_2 & p_3 & p_4 & p_5 & p_6 & w_1 \end{bmatrix} = \\ = 10^3 \cdot \begin{bmatrix} 0.017543717092354 & -0.025265281638022 & 0.040984285298812 & 1.855250747170489 & -3.397079720955738 & 6.251476924192442 & 0.013791203700439 \end{bmatrix},$$

resulting with the following optimal SF and SOF

$$K^{(0)} = 10^5 \cdot \begin{bmatrix} -1.014246236498231 & 0.000000000000000 & 0.000000000000000 & 0.708417204523523 \\ -0.177349433397692 & 0.000000000000000 & 0.000000000000000 & 0.124922842398310 \end{bmatrix}$$

$$K = 10^5 \cdot \begin{bmatrix} -1.014246236498231 & 0.708417204523523 \\ -0.177349433397692 & 0.124922842398310 \end{bmatrix},$$

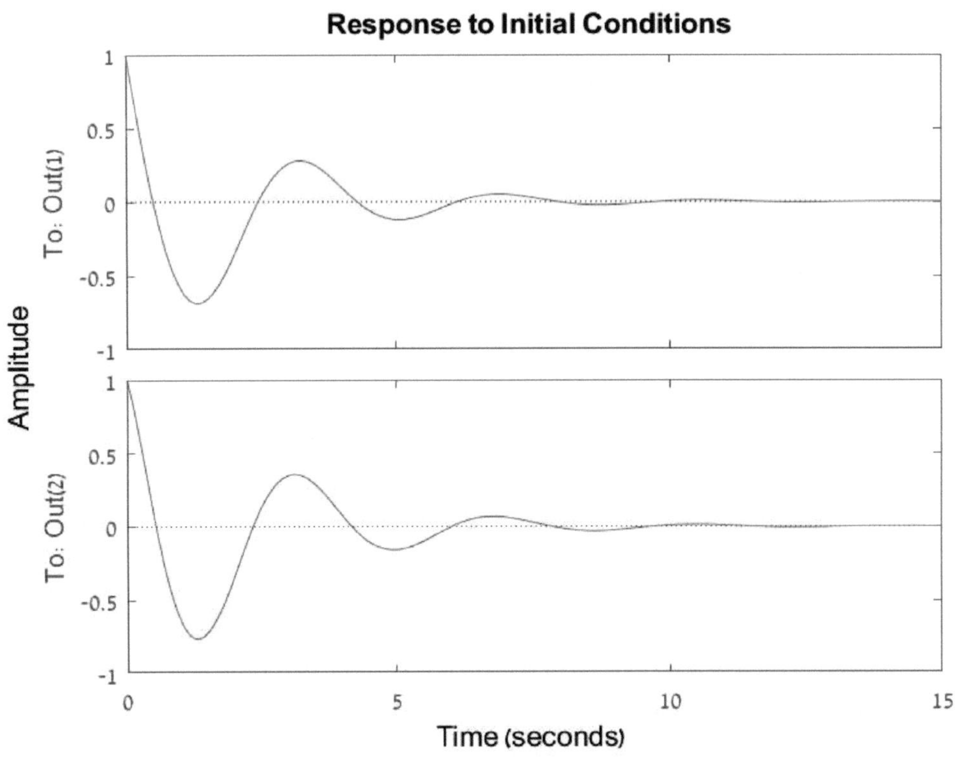

Figure 3.
Response of initial condition of the closed loop with the LQR SOF.

with $\|K\|_F = 1.256029021159584 \cdot 10^5$. The resulting closed-loop eigenvalues are:

$$\sigma\left(A^{(0)} - B^{(0)}KC^{(0)}\right) = \left\{ \begin{matrix} -9.682859336407926 \pm 0.940019732471932i \\ -0.157090195949127 \pm 1.083963060760387i \end{matrix} \right\}.$$

Figures 4 and **5** depict the impulse response and the step response of the closed loop with the pole-placement SOF. The amplitudes look reasonable but the settling time might be unsatisfactory.

Regarding the H_∞-norm of the closed loop, starting from:

$$\begin{bmatrix} p_1 & p_2 & p_3 & p_4 & p_5 & p_6 & w_1 \end{bmatrix} = \\ = \begin{bmatrix} 1 & 0 & 1 & 1 & 0 & 1 & 0 \end{bmatrix},$$

in $CPU - Time = 0.703125[sec]$ the fmincon function has converged to the optimal point

$$\begin{bmatrix} p_1 & p_2 & p_3 & p_4 & p_5 & p_6 & w_1 \end{bmatrix} = \\ = \begin{bmatrix} 0.888221316790683 & 0.005450463395221 & 0.509688006534611 & 0.351367770700493 & 0.108479534948988 & 2.135683618863295 & -0.023286321711901 \end{bmatrix},$$

resulting with the following optimal SF and SOF

$$K^{(0)} = 10^4 \cdot \begin{bmatrix} -0.434126348764817 & -0.000000000000000 & 0.000000000000000 & 6.230425140286776 \\ 6.139781209081431 & 0.000000000000000 & -0.000000000000000 & 0.417994561595739 \end{bmatrix}$$

$$K = 10^4 \cdot \begin{bmatrix} -0.434126348764817 & 6.230425140286776 \\ 6.139781209081431 & 0.417994561595739 \end{bmatrix},$$

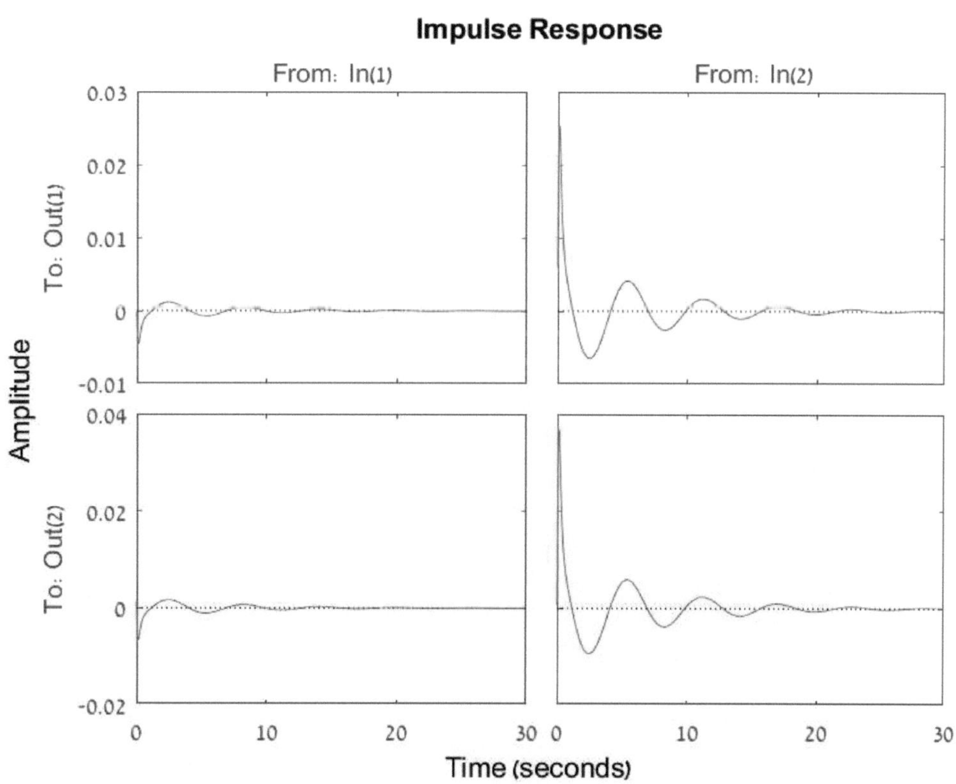

Figure 4.
Impulse response of the closed loop with the pole-placement SOF.

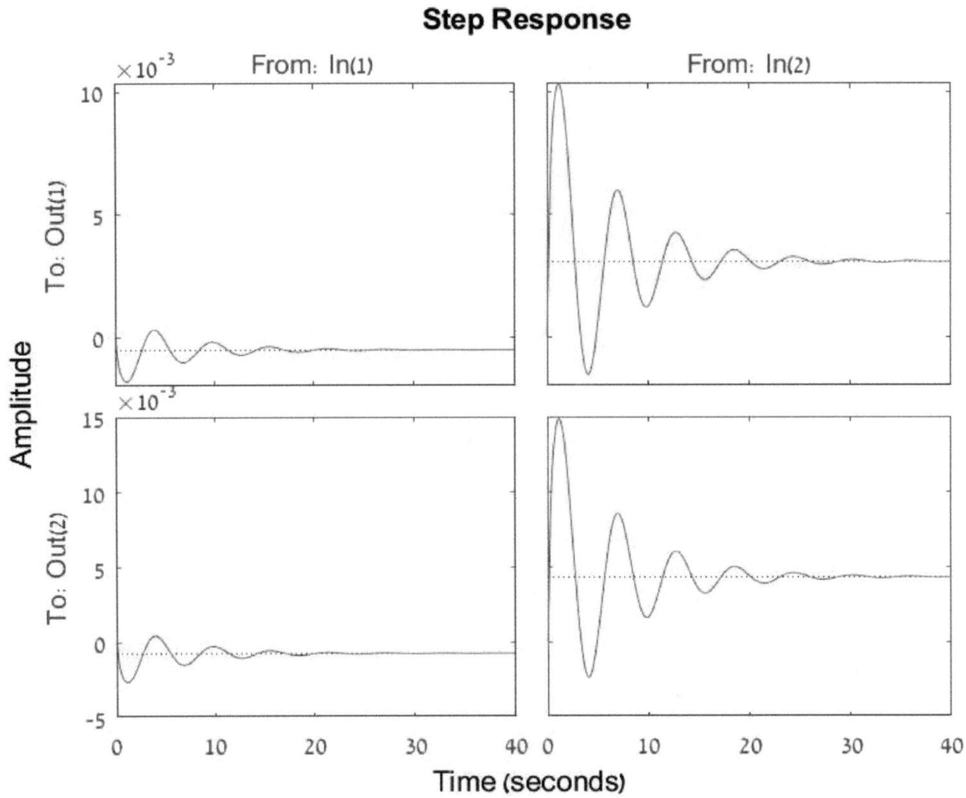

Figure 5.
Step response of the closed loop with the pole-placement SOF.

with $\|K\|_F = 8.768026908280017 \cdot 10^4$ and $\|T_{w,z}(s)\|_{H_\infty} = 1.631954397074613 \cdot 10^{-5}$. The resulting closed-loop eigenvalues are:

$$\sigma\left(A^{(0)} - B^{(0)}KC^{(0)}\right) = \left\{ \begin{array}{l} -356.9401845964764 \\ -90.9530886951882 \\ -1.0236129938218 \\ -0.5685837870690 \end{array} \right\}.$$

The simulation results of the closed-loop system are given in **Figure 6**, where w is normally distributed random disturbance, where each entry is $\mathcal{N}(0, 10^6)$ distributed. The maximum absolute values of the entries of $z = y$ are

$$\left[\begin{array}{ll} 0.034719714201842 & 0.014588756724050 \end{array}\right]^T,$$

and the maximum absolute values of the entries of x are

$$\left[\begin{array}{llll} 0.034719714201842 & 0.279876853192629 & 0.549124101316666 & 0.014588756724050 \end{array}\right]^T.$$

The results here are good.

Regarding the H_2-norm of the closed loop, starting from:

$$\begin{array}{l} \left[\begin{array}{ccccccc} p_1 & p_2 & p_3 & p_4 & p_5 & p_6 & w_1 \end{array}\right] = \\ = \left[\begin{array}{ccccccc} 1 & 0 & 1 & 1 & 0 & 1 & 0 \end{array}\right], \end{array}$$

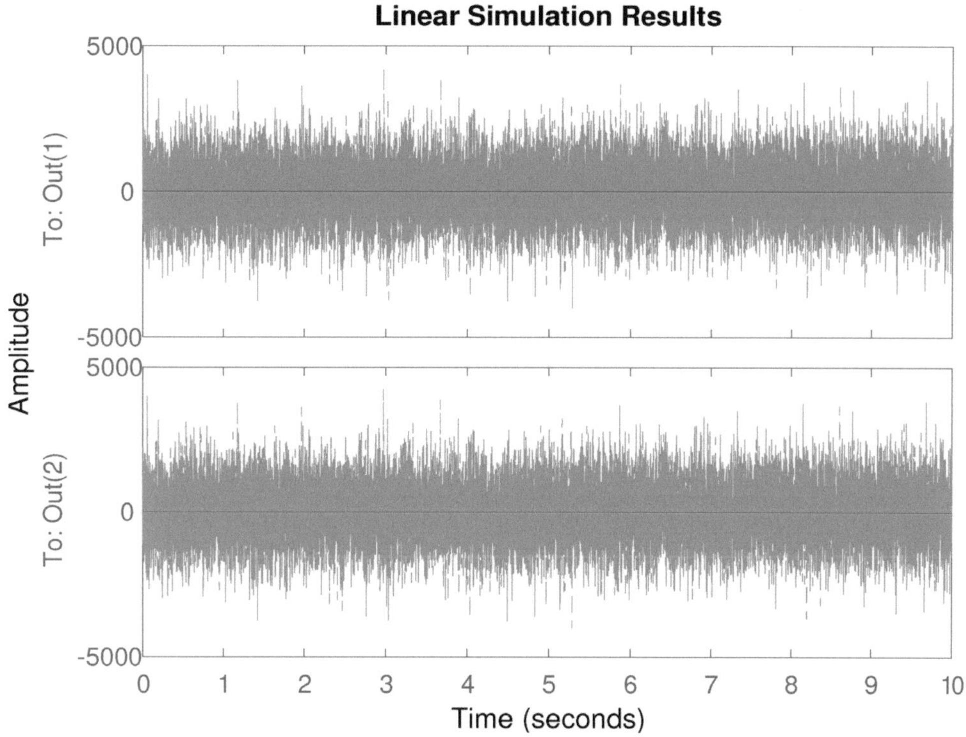

Figure 6.
Response of the closed loop with the optimal H_∞-norm SOF to a 10^6 variance zero-mean normally distributed random disturbance.

in $CPU - Time = 8.390625[sec]$ the fmincon function has converged to the optimal point

$$[p_1 \ p_2 \ p_3 \ p_4 \ p_5 \ p_6 \ w_1] =$$
$$= [0.891178477642138 \quad 0.006639774876451 \quad 0.508684007482598 \quad 0.000038288661546 \quad 0.000053652014908 \quad 0.000140441315775 \quad -0.021795268637180],$$

resulting with the following optimal SF and SOF

$$K^{(0)} = 10^9 \cdot \begin{bmatrix} 1.835262537587536 & 0.000000000003900 & 0.000000000012416 & 1.804218059606446 \\ 1.194244874676116 & -0.000000000002080 & 0.000000000007286 & 0.578593785387393 \end{bmatrix}$$
$$K = 10^9 \cdot \begin{bmatrix} 1.835262537587536 & 1.804218059606446 \\ 1.194244874676116 & 0.578593785387393 \end{bmatrix},$$

with $\|K\|_F = 2.895579903518702 \cdot 10^9$ and $\|T_{w,z}(s)\|_{H_2} = 2.289352128445973 \cdot 10^{-6}$. The resulting closed-loop eigenvalues are:

$$\sigma\left(A^{(0)} - B^{(0)}KC^{(0)}\right) = 10^6 \cdot \begin{Bmatrix} -8.825067937292802 \\ -1.087222799352728 \\ -0.000000561491544 \\ -0.000000982645227 \end{Bmatrix}.$$

The simulation results of the closed-loop system is given in **Figure 6**, where w is normally distributed random disturbance, where each entry is $\mathcal{N}(0, 10^6)$ distributed. The maximum absolute values of the entries of $z = y$ are

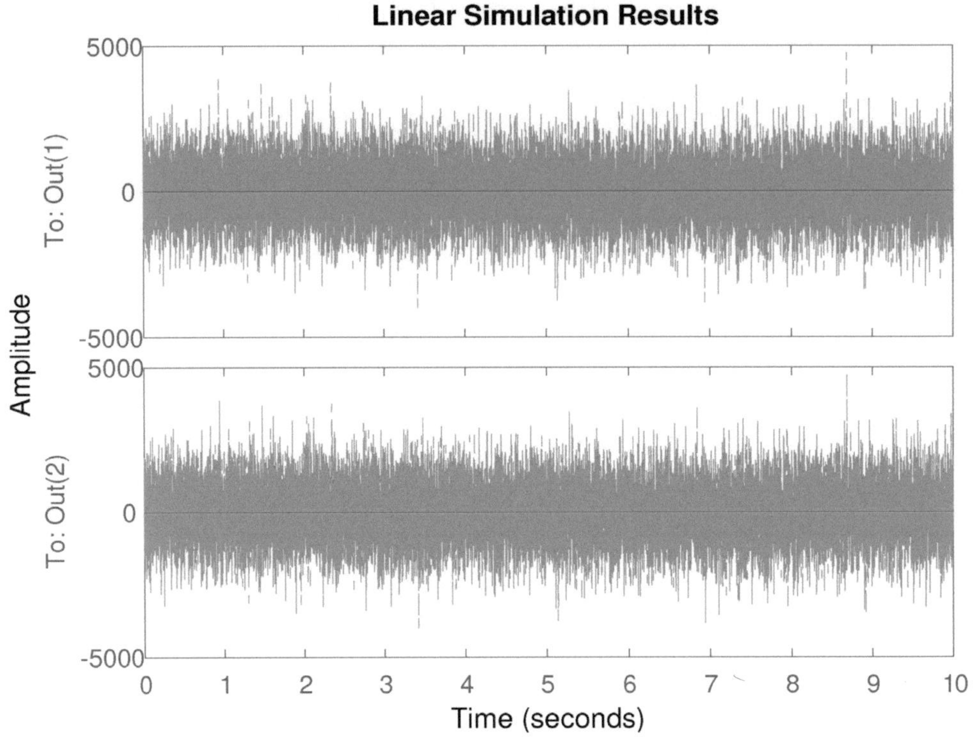

Figure 7.
Response of the closed loop with the optimal H_2-norm SOF to a 10^6 variance zero-mean normally distributed random disturbance.

$$10^{-5} \cdot [0.793706028985933 \quad 0.829879751812045]^T,$$

and the maximum absolute values of the entries of x are

$$10^{-5} \cdot [0.7937060289859 \quad 16.3502413073901 \quad 12.3720896708621 \quad 0.8298797518120]^T.$$

The results here are excellent.

We conclude that the best performance of the closed-loop system is achieved with the optimal H_2-norm SOF; however, since the Frobenius-norm of the SOF controller is high, the cost of construction and of operation of the SOF controller might be high, and there are no "free meals." Note also that by minimizing the SOF Frobenius-norm, the eigenvalues of the closed loop tend to get closer to the imaginary axes (to the region of lower degree of stability), while by minimizing the H_2-norm, the eigenvalues of the closed loop tend to escape from the imaginary axes (to the region of higher degree of stability). These are conflicting demands, and therefore, one should use some combination of the related key functions or to use some multiobjective optimization algorithm in order to get the best SOF in some or all of the needed key performance measures (**Figure** 7).

The following counterintuitive example shows that the SOF problem can be unsolvable (or hard to solve) even for small systems. In the example we show how nonexistence of SOF can be detected by the method:

Example 4.2 Let

$$A^{(0)} = \begin{bmatrix} 1 & 1 \\ 0 & 1 \end{bmatrix}, B^{(0)} = \begin{bmatrix} 1 \\ 1 \end{bmatrix}, C^{(0)} = [1 \quad 1].$$

Applying the algorithm we have

$$U^{(0)} = \frac{1}{\sqrt{2}}\begin{bmatrix} 1 & -1 \\ 1 & 1 \end{bmatrix}, A^{(1)} = \frac{1}{2}, B^{(1)} = -\frac{1}{2}.$$

The "while-loop" stops because $B^{(1)}B^{(1)+} = 1$. Let $P^{(1)} = p_1$ where $p_1 > 0$. Then,

$$K^{(1)} = B^{(1)+}\left(\frac{1}{2}\left(1 + A^{(1)}P^{(1)} + P^{(1)}A^{(1)T}\right)\right)\left(P^{(1)}\right)^{-1} = -\frac{(1+p_1)}{p_1}.$$

Let $\widehat{\Delta P^{(0)}}_{1,1} = p_2$, where $p_2 > 0$. Then

$$P^{(0)} = U^{(0)}\begin{bmatrix} \widehat{\Delta P^{(0)}}_{1,1} + K^{(1)}P^{(1)}K^{(1)T} & -K^{(1)}P^{(1)} \\ -P^{(1)}K^{(1)T} & P^{(1)} \end{bmatrix}U^{(0)T} =$$

$$= \frac{1}{2p_1}\begin{bmatrix} p_2p_1 + 1 & p_2p_1 + 1 + 2p_1 \\ p_2p_1 + 1 + 2p_1 & p_2p_1 + 1 + 4p_1 + 4p_1^2 \end{bmatrix}.$$

We therefore have:

$$K^{(0)} = B^{(0)+}\left(I_2 + A^{(0)}P^{(0)} + P^{(0)}A^{(0)T}\right)\left(I_2 - \frac{1}{2}B^{(0)}B^{(0)+}\right)\left(P^{(0)}\right)^{-1} =$$

$$= \frac{1}{4p_2p_1^3}\begin{bmatrix} 4p_2p_1^2 + 4p_1 + 6p_1^2 + 4p_1^3 + p_2^2p_1^2 + 2p_2p_1 + 4p_2p_1^3 + 1 \\ -2p_2p_1^2 - 2p_1 - 2p_1^2 - p_2^2p_1^2 - 2p_2p_1 - 1 + 4p_2p_1^3 \end{bmatrix}^T,$$

as the free parametrization of all the state feedbacks for which $A^{(0)} - B^{(0)}K^{(0)}$ is stable—for any choice of $p_1, p_2 > 0$. Now $L_{C^{(0)}} = \frac{1}{2}\begin{bmatrix} 1 & -1 \\ -1 & 1 \end{bmatrix}$ and the equations $K^{(0)}L_{C^{(0)}} = 0$ are equivalent to the single equation:

$$\frac{1}{4p_2p_1^3}\left(p_2^2p_1^2 + p_2\left(2p_1 + 3p_1^2\right) + \left(2p_1^3 + 4p_1^2 + 3p_1 + 1\right)\right) = 0.$$

Assuming $p_1, p_2 > 0$, the last equation implies that

$$p_2 = \frac{-\left(2p_1 + 3p_1^2\right) \pm \sqrt{\left(2p_1 + 3p_1^2\right)^2 - 4p_1^2\left(2p_1^3 + 4p_1^2 + 3p_1 + 1\right)}}{2p_1^2},$$

leading to a contradiction with p_2 being real positive number.

5. A parametrization for exact pole assignment via SFs

This section is based on the results reported in [34], where the proofs of the following lemma and theorem can be found. The aim of this section is to introduce a parametrization of all the SF's for the exact pole-assignment problem, when the set of eigenvalues can be given as free parameters (under some reasonable assumptions). This is done as part of the research of the problem of parametrization of all the SOFs for pole assignment. Note that the problem of exact pole assignment by SOFs is NP-hard (see [3]), meaning that an efficient algorithm for the problem probably does not exist, and therefore an effective description of the set of all solutions might not exist too. Also note that with SOFs, the feasible set Ω might exclude some open set from being a feasible set for the closed-loop spectrum (see [12]). These make the full aim very hard (if not impossible) to achieve. We therefore focus here on the problem of exact pole assignment via SFs.

Let the control system be given by:

$$\begin{cases} \Sigma(x(t)) = Ax(t) + Bu(t) \\ y(t) = Cx(t) \end{cases} \tag{35}$$

where $A \in \mathbb{R}^{n\times n}, B \in \mathbb{R}^{n\times m}, C \in \mathbb{R}^{r\times n}$, $\Sigma(x(t)) = \frac{d}{dt}x(t)$ in the continuous-time context and $\Sigma(x(t)) = x(t+1)$ in the discrete-time context. We assume without loss of generality that (A, B) is controllable. The problem of exact pole assignment by SF is defined as follows:

- (SF-EPA) Given a set $\Omega \subseteq \mathbb{C}_-$, $|\Omega| = n$ (in the discrete-time context $\Omega \subseteq \mathbb{D}_-$), symmetric with respect to the x-axis, find a state feedback $F \in \mathbb{R}^{m\times n}$ such that the closed-loop state-to-state matrix $E = A - BF$ has Ω as its complete set of eigenvalues, with their given multiplicities.

In [13], a closed form of all the exact pole-placement SFs is proved (up to a set of measure 0), based on Moore's method. In order to minimize the inaccuracy of the eigenvalues final placement and in order to minimize the Frobenius-norm of the feedback, a convex combination of the condition number of the similarity matrix and of the feedback norm was minimized. The parametrization proposed in [13] is based on the assumptions that there exists at least one real state feedback that leads to a diagonalizable state-to-state closed-loop matrix and that B is full rank. A necessary condition for such SF to exist is that the final multiplicity of any eigenvalue is less than or equal to $rank(B)$. Here, we do not assume that B is full rank and we only assume that Ω contains sufficient number of real eigenvalues. A survey of most of the methods for robust pole assignment via SFs or by SOFs and the formulation of these methods as optimization problems with optimality necessary conditions is given in [44]. In [45] a performance comparison of most of the algorithmic methods for robust pole placement is given. A formulation of the general problem of robust exact pole assignment via SFs as an SDP problem and LMI-based linearization is introduced in [46], where the robustness is with respect to the condition number of the similarity matrix, which is made in order to hopefully minimize the inaccuracy of the eigenvalues final placement. Unfortunately, one probably cannot gain a parametric closed form of the SFs from such formulations. Moreover, the following proposed method is exact and therefore enables the use of the parametrization free parameters for other (and maybe more important) optimization purposes. Note that since the proposed method is exact, the closed-loop eigenvalues thyself can be inserted to the problem as parameters.

A completely different notion of robustness with respect to pole placement is considered in the following works:

Robust pole placement in LMI regions and H_∞ design with pole placement in LMI regions are considered in [47, 48], respectively. An algorithm based on alternating projections is introduced in [15], which aims to solve efficiently the problem of pole placement via SOFs. A randomized algorithm for pole placement via SOFs with minimal norm, in nonconvex or unconnected regions, is considered in [20].

Let $\Omega = \left\{ \underbrace{\alpha_1, \overline{\alpha_1}}_{c_1 \text{ times}}, \dots, \underbrace{\alpha_m, \overline{\alpha_m}}_{c_m \text{ times}}, \underbrace{\beta_1}_{r_1 \text{ times}}, \dots, \underbrace{\beta_\ell}_{r_\ell \text{ times}} \right\}$, be the intended closed-loop eigenvalues, where the αs denote the paired complex-conjugate eigenvalues (with nonzero imaginary part), the βs denote the real eigenvalues, and $2c_1, \dots, 2c_m, r_1, \dots, r_\ell$ denote their respective multiplicities, where $2\sum_{i=1}^{m} c_i + \sum_{j=1}^{\ell} r_j = n$. In the following we would say that the size of the set (actually, the multiset) Ω is n (counting multiplicities) and we would write $|\Omega| = n$. Note that (A, B) is controllable if and only if (A, BB^+) is controllable, and also note that BB^+ is a real symmetric matrix with simple eigenvalues in the set $\{0, 1\}$ and thus is orthogonally diagonalizable matrix. Let U denote an orthogonal matrix such that:

$$\widehat{B} = U^T BB^+ U = \begin{bmatrix} I_k & 0 \\ 0 & 0 \end{bmatrix} = bdiag(I_k, 0), \tag{36}$$

where $k = rank(B) = rank(BB^+) \geq 1$ since (A, B) is controllable, and let $\widehat{A} = U^T A U = \begin{bmatrix} \widehat{A}_{1,1} & \widehat{A}_{1,2} \\ \widehat{A}_{2,1} & \widehat{A}_{2,2} \end{bmatrix}$ be partitioned accordingly. We cite here the following lemma taken from [34] connecting between the controllability of the given system and the controllability of its sub-system:

Lemma 5.1 In the notations above, (A, BB^+) is controllable if and only if $\left(\widehat{A}_{2,2}, \widehat{A}_{2,1}\widehat{A}_{2,1}^{+}\right)$ is controllable.

Again, we use the recursive controllable structure. Let $U^{(0)} = U$ and let $A^{(0)} = A, B^{(0)} = B, n_0 = n, k_0 = rank\left(B^{(0)}\right)$. Similarly, let $U^{(1)}$ be an orthogonal matrix such that $U^{(1)T} B^{(1)} B^{(1)+} U^{(1)} = bdiag(I_{k_1}, 0)$, where $B^{(1)} = \widehat{A}_{2,1}$. Let $A^{(1)} = \widehat{A}_{2,2}, n_1 = n_0 - k_0, k_1 = rank\left(B^{(1)}\right)$. Now, Lemma 5.1 implies that $\left(A^{(1)}, B^{(1)}\right)$ is controllable since $\left(A^{(0)}, B^{(0)}\right)$ is controllable. Recursively, assume that the pair $\left(A^{(i)}, B^{(i)}\right)$ is controllable. Let $U^{(i)}$ be an orthogonal matrix such that $\widehat{B^{(i)}} = U^{(i)T} B^{(i)} B^{(i)+} U^{(i)} = bdiag(I_{k_i}, 0)$, where $k_i \geq 1$ (since $\left(A^{(i)}, B^{(i)}\right)$ is controllable). Let $\widehat{A^{(i)}} = U^{(i)T} A^{(i)} U^{(i)} = \begin{bmatrix} \widehat{A^{(i)}}_{1,1} & \widehat{A^{(i)}}_{1,2} \\ \widehat{A^{(i)}}_{2,1} & \widehat{A^{(i)}}_{2,2} \end{bmatrix}$ be partitioned accordingly, with sizes $k_i \times k_i$ and $(n_i - k_i) \times (n_i - k_i)$ of the main block-diagonal blocks. Let $A^{(i+1)} = \widehat{A^{(i)}}_{2,2}, B^{(i+1)} = \widehat{A^{(i)}}_{2,1}, n_{i+1} = n_i - k_i, k_i = rank\left(B^{(i)}\right)$. Then, Lemma 5.1 implies that $\left(A^{(i+1)}, B^{(i+1)}\right)$ is controllable. The recursion stops when $B^{(i)} B^{(i)+} = I_{k_i}$ for some $i = b$ (which we call the base case). Note that in the worst case, the recursion stops when the rank $k_b = 1$.

Theorem 5.1 In the above notations, assume that $\sum_{j=1}^{\ell} r_j \geq a$, where a is the number of parity alternations in the sequence $\langle n_0, n_1, \dots, n_b \rangle$. Let $\Omega_0 = \Omega$. Then, there exist a sequence $\Omega_0 \supseteq \Omega_1 \supseteq \cdots \supseteq \Omega_b$ of symmetric sets with size $|\Omega_i| = n_i$ (counting multiplicities) and there exist a real state feedback $F_i = F_i(G_{i+1}, F_{i+1})$ such that $\sigma\left(A^{(i)} - B^{(i)} F_i\right) = \Omega_i$. Moreover, an explicit (recursive) formula for $F_i(G_{i+1}, F_{i+1})$ is given by:

$$\begin{cases} F_i = B^{(i)+}W^{(i)} \\ W^{(i)} = U^{(i)}\widehat{W}^{(i)}U^{(i)T} \\ \widehat{W}^{(i)} = \begin{bmatrix} \widehat{W}^{(i)}_{1,1} & \widehat{W}^{(i)}_{1,2} \\ 0 & 0 \end{bmatrix} \\ \widehat{W}^{(i)}_{1,1} = \widehat{A}^{(i)}_{1,1} + F_{i+1}\widehat{A}^{(i)}_{2,1} - G_{i+1} \\ \widehat{W}^{(i)}_{1,2} = \widehat{A}^{(i)}_{1,2} + F_{i+1}\widehat{A}^{(i)}_{2,2} - G_{i+1}F_{i+1}, \end{cases} \tag{37}$$

where $\sigma\left(\widehat{A}^{(i)}_{2,2} - \widehat{A}^{(i)}_{2,1}F_{i+1}\right) = \sigma\left(A^{(i+1)} - B^{(i+1)}F_{i+1}\right) = \Omega_{i+1}$ and G_{i+1} is arbitrary real matrix such that $\sigma(G_{i+1}) = \Omega_i \backslash \Omega_{i+1}$.

Example 5.1 Consider the problem of exact pole assignment via SF for the same system from Example 4.1. We therefore assume here that the full state is available for feedback control. Now, using the calculations from Example 4.1, we have: $\langle n_0, n_1 \rangle = \langle 4, 2 \rangle$ implying that the number of parity alternations is $a = 0$. We therefore can assign by the method any symmetric set of eigenvalues to the closed loop. Let $\Omega_0 = \{\alpha, \overline{\alpha}, \beta, \overline{\beta}\}$ be the eigenvalues to be assigned, and let $\Omega_1 = \{\beta, \overline{\beta}\}$. Now,

$$F_1 = B^{(1)+}\left(A^{(1)} - G_2\right),$$

where

$$G_2 = \begin{bmatrix} \Re(\beta) & \Im(\beta) \\ -\Im(\beta) & \Re(\beta) \end{bmatrix},$$

and

$$F_0 = B^{(0)+}W^{(0)},$$

where

$$\begin{aligned} W^{(0)} &= U^{(0)}\widehat{W^{(0)}}U^{(0)T} \\ \widehat{W^{(0)}} &= \begin{bmatrix} \widehat{W^{(0)}}_{1,1} & \widehat{W^{(0)}}_{1,2} \\ 0_2 & 0_2 \end{bmatrix} \\ \widehat{W^{(0)}}_{1,1} &= \widehat{A^{(0)}}_{1,1} + F_1\widehat{A^{(0)}}_{2,1} - G_1 \\ \widehat{W^{(0)}}_{1,2} &= \widehat{A^{(0)}}_{1,2} + F_1\widehat{A^{(0)}}_{2,2} - G_1F_1 \\ G_1 &= \begin{bmatrix} \Re(\alpha) & \Im(\alpha) \\ -\Im(\alpha) & \Re(\alpha) \end{bmatrix}. \end{aligned}$$

We have completed the pole-assignment SF parametrization. As an application, assume that $\alpha = -10 + i, \beta = -1 + 0.1i$. Then,

$$F_0 = 10^3 \cdot \begin{bmatrix} -2.312944765727539 & 0.063274421635669 & -0.033598570868307 & 7.136847864481691 \\ 2.382051359668675 & -0.002249112006026 & 0.011460219946760 & 0.432788287638212 \end{bmatrix},$$

resulting with the closed-loop eigenvalues:

$$\sigma\left(A^{(0)} - B^{(0)}F_0\right) = \left\{ \begin{array}{c} -10.000000000000000 \pm 1.000000000000004i \\ -1.000000000000003 \pm 0.099999999999998i \end{array} \right\}.$$

In our calculations, we used MATLAB®, which has a general precession of $5 - 7$ significant digits in computing eigenvalues. Thus, we have almost no loss of digits by the method. For a comparison, see the last case in Example 4.1 and note that while exact pole assignment can be achieved by SF, in general, it cannot be achieved by SOF because the last is a NP-hard problem (see the introduction of this section). Even regional pole placement is hard to achieve by SOF because of the nonconvexity of the SOF feasibility domain.

Remark 5.1 Note that the indices $\langle k_0, k_1, \ldots, k_b \rangle$ as well as the indices $\langle n_0, n_1, \ldots, n_b \rangle$, can be calculated from (A, B) in advance. After calculating these indices and the number a of parity alternations in the sequence $\langle n_0, n_1, \ldots, n_b \rangle$, the designer can define Ω as to satisfy the assumption of Theorem 5.1, i.e., being symmetric with at least a real eigenvalues, in a parametric way, and get a parametrization of all the real SF leading to Ω as the set of closed-loop eigenvalues. Next, the designer can play with the specific values of these and of other free parameters, in order to gain the needed closed-loop performance requirements. This is in contrast with other methods where the parametrization is calculated ad-hoc for a specific set of eigenvalues, where any change in the set of eigenvalues necessitates new execution of the method.

Remark 5.2 Note that F_{i+1} can be replaced by $F_{i+1} + \left(I - B^{(i)+}B^{(i)}\right)H_{i+1}$ where H_{i+1} is any real matrix, without changing the closed-loop eigenvalues (if one seek feedbacks with minimal Frobenius-norm, then he should take $H_{i+1} = 0$, otherwise he should leave H_{i+1} as another free parameter). Thus, the freeness in $\langle H_b, \ldots, H_1, H_0 \rangle$ and in $\langle G_{b+1}, G_b, \ldots, G_1 \rangle$ makes the freeness in F_0 (e.g. in order to globally optimize the H_∞-norm of the closed loop, the H_2-norm or the LQR functional of the closed loop or any other performance key thereof). Note also that the sequences $\langle F_b, \ldots, F_1, F_0 \rangle$ and $\langle G_{b+1}, G_b, \ldots, G_1 \rangle$ can be calculated for Ω as in Theorem 5.1, where the eigenvalues in Ω are given as free parameters. In that case, it can be easily proved by induction that the state feedbacks $\langle F_b, \ldots, F_1, F_0 \rangle$ depend polynomially on the eigenvalues parameters and on the other free parameters mentioned above (for complex eigenvalue α they depend polynomially on $\Re(\alpha), \Im(\alpha)$).

Finally, it is wort mentioning that the complementary theorem of Theorem 5.1 was also proved in [34], meaning that under the assumptions of Theorem 5.1, any SF that solves the problem has the form given in the theorem (up to a factor of the form given in Remark 5.2).

6. Concluding remarks

In this chapter, we have introduced an explicit free parametrization of all the stabilizing SF's of a controllable pair (A, B). This enables global optimization over the set of all the stabilizing SF's of such pair, because the parametrization is free. For a system triplet (A, B, C), we have shown how to get the parametrization of all the SOFs of the system by parameterizing all the SFs of (A, B) and all the SFs of $\left(A^T, C^T\right)$ and then imposing the compatibility constraint (28). We have also shown a parametrization of all the SOFs of the system triplet (A, B, C) by imposing the linear constraint $K^{(0)}L_{C^{(0)}} = 0$ on the SF $K^{(0)}$ of the pair (A, B), where $K^{(0)}$ was defined recursively and parameterizes the set of all SFs of (A, B). This leads to a set of polynomial equations (after multiplying by the l.c.m. of the denominators of the rational entries of $K^{(0)}$) and inequalities that can be brought to polynomial equations. The resulting polynomial set of equations can be solved (parametrically) by using the Gröbner basis method (see e.g., [49–52]). By applying the Gröbner basis method, one would get an indication to the existence of solutions and in case that solutions do exist, it would tell what are the free parameters and how other parameters depend on the free parameters. It seems that the proposed method makes the Gröbner basis computations overhead (or other methods thereof) reduced significantly, thus enabling SOF global optimization for larger systems.

In view of Theorem 5.1 (with its complementary theorem proved in [34]), we have introduced a sound and complete parametrization of all the state feedbacks F which make the matrix $\widehat{E} = U^T(A - BF)U$ k-complementary $(n - k)$-invariant with respect to Ω_1 (see [34] for the definition and properties), where U is orthogonal such that $U^T BB^+ U = bdiag(I_k, 0), k = rank(B)$, where Ω is symmetric and has at least a (being the parity alternations in the sequence $\langle n_0, \dots, n_b \rangle$) real eigenvalues, where $\Omega_1 \subseteq \Omega$ is symmetric with maximum real eigenvalues with size $|\Omega_1| = n - k$. Assuming Ω as above, we have generalized the results of [13] in the sense that we do not assume the existence of real state feedback F that brings the closed-loop $E = A - BF$ to diagonalizable matrix, which actually means that the geometric and algebraic multiplicity coincides for any eigenvalue of the closed loop, and we do not assume the restriction on the multiplicity of each eigenvalue to be less than or equal to $rank(B)$. However, in cases where the number of real eigenvalues in Ω is less than a, one should use the parametrizations given in [13], in [45] or in the references there. Note that in communication systems, where complex SFs and SOFs are sought, the introduced method is complete (with no restrictions) since the number of parity alternations and the restriction on Ω to contain as much real eigenvalues, were needed only to guarantee that F_i for $i = b, \dots, 0$ is real in each stage, which is needless in communication systems.

In view of Example 5.1, one can see that the accuracy of the final location of the closed-loop eigenvalues given by the proposed method depends only on the accuracy of computing $B^{(i)+}$ and $U^{(i)}$ for $i = 0, \dots, b$ and in the algorithm that we have to compute the closed-loop eigenvalues (see [53], for example) in order to validate their final location, and it has nothing to do with the specific values of the specific eigenvalues given in Ω. Therefore, by the proposed method, once that $B^{(i)+}$ and $U^{(i)}$ for $i = 0, \dots, b$ were computed as accurate as possible, the location of the closed-loop eigenvalues will be accurate accordingly. Thus, by the proposed method the designer can save time since he can do it parametrically only once, and afterward he only needs to play with the specific values of the eigenvalues until he gets a satisfactory closed-loop performance, where he can be sure that the accuracy of the final placement will be the same for all of his trials independently on the specific values of the chosen eigenvalues. Also, the given parametrization of F_0 is polynomially dependent on the free parameters and thus is very convenient for applying automatic differentiation and optimization methods.

To conclude, we have introduced parametrizations of SFs and SOFs that are based on the recursive controllable structure that was discovered in [35]. The results has powerful implications for real-life systems, and we expect for more results in this direction. Unfortunately, for uncertain systems, the method cannot work directly because of the dependencies of (A, B, C) in uncertain parameters, for which we cannot compute $U^{(i)}$ for $i = 0, \dots, b$. However, if a nominal system $(\tilde{A}, \tilde{B}, \tilde{C})$ is known accurately then, the method can be applied to that system and the free parameters of the parametrization can be used to "catch" the uncertainty of the whole system, together with the closed-loop performance requirements. The research of this method will be left for a future work.

Author details

Yossi Peretz
Department of Computer Sciences, Lev Academic Center, Jerusalem College of Technology, Jerusalem, Israel

*Address all correspondence to: yosip@g.jct.ac.il

References

[1] Nemirovskii A. Several NP-hard problems arising in robust stability analysis. Mathematics of Control, Signals, and Systems. Springer. 1993;**6** (2):99-105

[2] Blondel V, Tsitsiklis JN. NP-hardness of some linear control design problems. SIAM Journal on Control and Optimization. 1997;**35**(6):2118-2127

[3] Fu M. Pole placement via static output feedback is NP-hard. IEEE Transactions on Automatic Control. 2004;**49**(5):855-857

[4] Peretz Y. On applications of the ray-shooting method for structured and structured-sparse static-output-feedbacks. International Journal of Systems Science. 2017;**48**(9):1902-1913

[5] Kučera V, Trofino-Neto A. Stabilization via static output feedback. IEEE Transactions on Automatic Control. 1993;**38**(5):764-765

[6] de Souza CC, Geromel JC, Skelton RE. Static output feedback controllers: Stability and convexity. IEEE Transactions On Automatic Control. 1998;**43**(1):120-125

[7] Cao YY, Lam J, Sun YX. Static output feedback stabilization: An ILMI approch. Automatica. 1998;**34**(12):1641-1645

[8] Cao YY, Sun YX, Lam J. Simultaneous stabilization via static output feedback and state feedback. IEEE Transactions on Automatic Control. 1998;**44**(6):1277-1282

[9] Quoc TD, Gumussoy S, Michiels W, Diehl M. Combining convex-concave decompositions and linearization approaches for solving BMIs, with application to static output feedback. IEEE Transactions on Automatic Control. 2012;**57**(6): 1377-1390

[10] Mesbahi M. A semi-definite programming solution of the least order dynamic output feedback synthesis problem. Procedings of the 38'th IEEE Conference on Decision and Control. 1999;(2):1851-1856

[11] Henrion D, Loefberg J, Kočvara M, Stingl M. Solving Polynomial static output feedback problems with PENBMI. In: Proc. Joint IEEE Conf. Decision Control and Europ. Control Conf.; 2005; Sevilla, Spain. 2005

[12] Eremenko A, Gabrielov A. Pole placement by static output feedback for generic linear systems. SIAM Journal on Control and Optimization. 2002;**41**(1): 303-312

[13] Schmid R, Pandey A, Nguyen T. Robust pole placement with Moore's algorithm. IEEE Transactions on Automatic Control. 2014;**59**(2):500-505

[14] Fazel M, Hindi H, Boyd S. Rank minimization and applications in system theory. In: Proceedings of the American Control Conference. 2004. pp. 3273-3278

[15] Yang K, Orsi R. Generalized pole placement via static output feedback: A methodology based on projections. Automatica. 2006;**42**:2143-2150

[16] Vidyasagar M, Blondel VD. Probabilistic solutions to some NP-hard matrix problems. Automatica. 2001;**37**: 1397-1405

[17] Tempo R, Calafiore G, Dabbene F. Randomized Algorithms for Analysis and Control of Uncertain Systems. London: Springer-Verlag; 2005

[18] Tempo R, Ishii H. Monte Carlo and Las Vegas randomized algorithms for systems and control. European Journal of Control. 2007;**13**:189-203

[19] Arzelier D, Gryazina EN, Peaucelle D, Polyak BT. Mixed LMI/randomized fcmethods for static output feedback control. In: Proceedings of the American Control Conference, IEEE Conference Publications. 2010. pp. 4683-4688

[20] Peretz Y. A randomized approximation algorithm for the minimal-norm static-output-feedback problem. Automatica. 2016;**63**:221-234

[21] Apkarian P, Noll D. Nonsmooth H_∞ synthesis. IEEE Transactions on Automatic Control. 2006;**51**(1):71-86

[22] Burke JV, Lewis AS, Overton ML. Stabilization via nonsmooth, nonconvex optimization. IEEE Transactions On Automatic Control. 2006;**51**(11): 1760-1769

[23] Gumussoy S, Henrion D, Millstone M, Overton ML. Multiobjective robust control with HIFOO 2.0. In: Proceedings of the IFAC Symposium on Robust Control Design; 2009; Haifa, Israel. 2009

[24] Borges RA, Calliero TR, Oliveira CLF, Peres PLD. Improved conditions for reduced-order H_∞ filter design as a static output feedback problem. In: American Control Conference, San-Francisco, CA, USA. 2011

[25] Peretz Y. On application of the ray-shooting method for LQR via static-output-feedback. MDPI Algorithms Journal. 2018;**11**(1):1-13

[26] Zheng F, Wang QG, Lee TH. On the design of multivariable PID controllers via LMI approach. Automatica. 2002;**38**: 517-526

[27] Peretz Y. A randomized algorithm for optimal PID controllers. MDPI Algorithms Journal. 2018;**11**(81):1-15

[28] Lin F, Farad M, Jovanović M. Sparse feedback synthesis via the alternating direction method of multipliers. In: IEEE, Proceedings of the American Control Conference (ACC), 2012. pp. 4765-4770

[29] Lin F, Farad M, Jovanović M. Augmented lagrangian approach to design of strutured optimal state feedback gains. IEEE Transactions On Automatic Control. 2011;**56**(12): 2923-2929

[30] Gillis N, Sharma P. Minimal-norm static feedbacks using dissipative Hamiltonian matrices. Linear Algebra and its Applications. 2021;**623**:258-281

[31] Silva RN, Frezzatto L. A new parametrization for static output feedback control of LPV discrete-time systems. Automatica. 2021;**128**:109566

[32] de Oliveira AM, Costa OLV. On the H_2 static output feedback control for hidden Markov jump linear systems. Annals of the Academy of Romanian. Scientists. Series on Mathematics and its Applications. 2020;**12**(1-2)

[33] Ren D, Xiong J, Ho DW. Static output feedback negative imaginary controller synthesis with an H_∞ norm bound. Automatica. 2021;**126**:109157

[34] Peretz Y. On parametrization of all the exact pole-assignment state feedbacks for LTI systems. IEEE Transactions on Automatic Control. 2017;**62**(7):3436-3441

[35] Peretz Y. A characterization of all the static stabilizing controllers for LTI systems. Linear Algebra and its Applications. 2012;**437**(2):525-548

[36] Iwasaki T, Skelton RE. All controllers for the general H_∞ control problem: LMI existence conditions and state space formulas. Automatica. 1994; **30**(8):1307-1317

[37] Iwasaki T, Skelton RE. Parametrization of all stabilizing controllers via quadratic lyapunov

functions. Journal of Optimization Theory and Applications. 1995;**85**(2): 291-307

[38] Skelton RE, Iwasaki T, Grigoriadis KM. A Unified Algebraic Approach To Linear Control Design. London: Tylor & Francis Ltd.; 1998

[39] Piziak R, Odell PL. In: Nashed Z, Taft E, editors. Matrix Theory: From Generalized Inverses to Jordan Form. Chapman & Hall/CRC \& Francis Group, 6000 Broken Sound Parkway NW, Suit 300, Boca Raton, FL 33487-2742; 2007. p. 288

[40] Karlheinz S. Abstract Algebra With Applications. Marcel Dekker, Inc.; 1994

[41] Ohara A, Kitamori T. Geometric structures of stable state feedback systems. IEEE Transactions on Automatic Control. 1993;**38**(10):1579-1583

[42] Leibfritz F. COMPleib: Constrained matrix-optimization problem library—A collection of test examples for nonlinear semidefinite programs, control system design and related problems. In: Dept. Math., Univ. Trier, Germany, Tech.-Report, (2003).

[43] Tadashi I, Hai-Jiao G, Hiroshi T. A design of discrete-time integral controllers with computation delays via loop transfer recovery. Automatica. 1992;**28**(3):599-603

[44] Chu EK. Optimization and pole assignment in control system design. International Journal of Applied Mathematics and Computer Science. 2001;**11**(5):1035-1053

[45] Pandey A, Schmid R, Nguyen T, Yang Y, Sima V, Tits AL. Performance survey of robust pole placement methods. In: IEEE 53rd Conference on Decision and Control; December 15-17; Los Angeles, California, USA. 2014. pp. 3186-3191

[46] Ait Rami M, El Faiz S, Benzaouia A, Tadeo F. Robust exact pole placement via an LMI-based algorithm. IEEE Transactions on Automatic Control. 2009;**54**(2):394-398

[47] Chilali M, Gahinet P. H_∞ design with pole placement constraints: An LMI approach. IEEE Transactions on Automatic Control. 1996;**41**(3):358-367

[48] Chilali M, Gahinet P, Apkarian P. Robust pole placement in LMI regions. IEEE Transactions on Automatic Control. 1999;**44**(12):2257-2270

[49] Buchberger B. Gröbner bases and system theory. Multidimentional Systems and Signal Processing. 2001;**12**: 223-251

[50] Shin HS, Lall S. Optimal decentralized control of linear systems via Groebner bases and variable elimination. In: American Control Conference. Baltimore, MD, USA: Marriott Waterfront; 2010. pp. 5608-5613

[51] Lin Z. Gröbner bases and applications in control and systems. In: IEEE, 7th International Conference On Control, Automation, Robotics And Vision (ICARCV'02). 2002. pp. 1077-1082

[52] Lin Z, Xu L, Bose NK. A tutorial on Gröbner bases with applications in signals and systems. IEEE Transactions on Circuits and Systems. 2008;**55**(1): 445-461

[53] Pan VY. Univariate polynomials: nearly optimal algorithms for numerical factorization and root-finding. Elsevier Journal of Symbolic Computation. 2002; **33**(5):701-733

Chapter 6

Experimental Studies of Asynchronous Electric Drives with "Stepwise" Changes in the Active Load

Vladimir L. Kodkin, Alexandr S. Anikin,
Alexandr A. Baldenkov and Natalia A. Loginova

Abstract

The article offers the results of experimental studies of asynchronous electric 10 motors with "squirrel cage" rotor with frequency control. The results of bench tests of the modes of parrying stepwise changes in the load created by a similar frequency-controlled electric drive are presented. A preliminary qualitative analysis of the known control methods is carried out and it is shown that the assumptions made when creating their algorithms in the modes of countering the load become too significant. The reasons for this are the fundamental inaccuracies of the vector equations of asynchronous electric motors with frequency regulation. The proposed interpretation of asynchronous electric motors by nonlinear continuous transfer functions, outlined in the articles written by the same authors earlier, and the corrections they proposed turned out to be more accurate for the operating modes under consideration than the traditional methods of interpretation and correction of the frequency control of asynchronous electric motors This made it possible to assess as objectively as possible the effectiveness of the interpretation of asynchronous electric drives and methods of their regulation. Numerous articles on this topic over the past 25–30 years have not provided such results.

Keywords: asynchronous drive, frequency regulation, dynamic positive feedbacks, active stator current, rotor current, signal spectrum

1. Introduction

The paradoxical situation has developed in the last 20 years in the frequency control asynchronous electric drives.

On the one hand, the frequency control of squirrel cage induction motor (SCIM) with semiconductor frequency and voltage converters (FC) is widespread in industry and energy, has several universal control methods that provide applied in increasingly complex and accurate technological units [1–4].

On the other hand, there are currently several fundamental unresolved theoretical problems that have been formed more than 100 years ago—when forming the theory of AC electrical machines.

The description of the processes of AC electric drives control processes is the vector equations and flowing from them—substitution schemes and vector diagrams (**Figure 1**) ([1], p. 18):

$$
\begin{aligned}
u_1 &= i_1\left(r_1 + jx_{1\sigma}\right) + ji_m x_m \\
0 &= i_2\left(\frac{r_2}{s} + jx_{2\sigma}\right) + ji_m + x_m \\
m &= \frac{m_1}{2} Z_P L_m |i_2 \times i_m| = \frac{m_1}{2} L_m I_{2max} I_{m\ max} \sin\psi \\
x_1 &= \omega_1 L_1; x_2 = \omega_1 L_2; x_m = \omega_1 L_m
\end{aligned}
\tag{1}
$$

These equations have a number of assumptions and simplifications acceptable to static modes, but completely erroneous for dynamic.

First, these equations suggest the sinusoidal nature of currents and stresses formed in asynchronous electric motors. The theory of control that operates vectors is simply not able to take into account any more components of these variables. At the same time, the presence of such components—the so-called "higher harmonics" in the currents of the motors and FC recognize all experts. However, in the engine equation, these components are not included, and only electrical interference is taken into account from all problems.

Secondly, the operating vectors instead of sinusoidal functions, interpreting currents in the stator and the rotor of the engines are valid only if the frequencies of their change are constant. Only in this case the differential equations "pass" in the vector and can be significantly simplified. It is important to note that even evaluating the error of such a replacement during frequency variations is analytically very difficult.

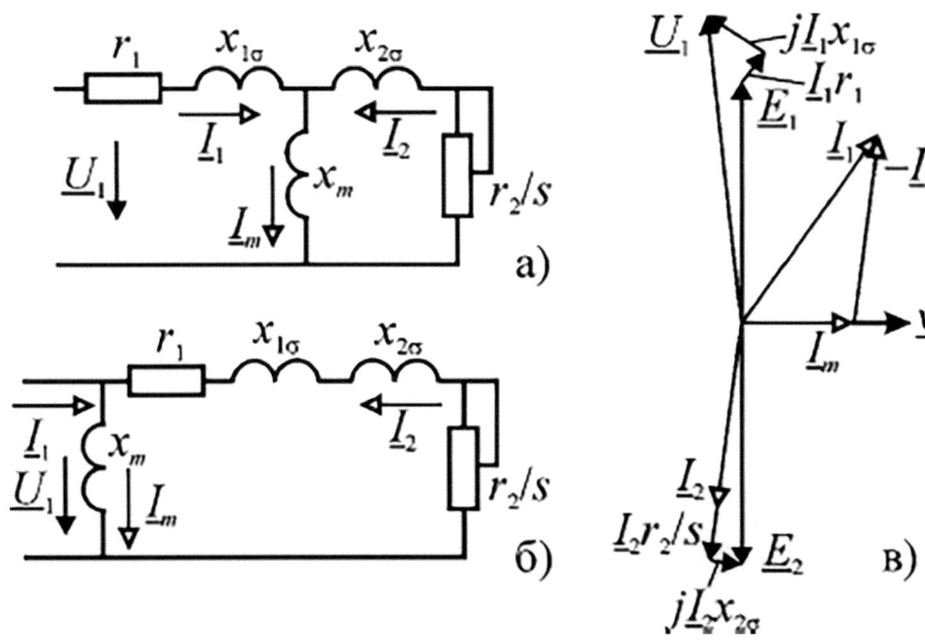

Figure 1.
Substitution scheme, vector diagram of asynchronous motor and vector equations in traditional form [1]. X_1, X_2 *– inductive resistances stator and rotor;* r_1, r_2 *– active resistances stator and rotor;* I_1, I_2 *– stator and rotor currents; and* E_2, U_1 *– rotor EMF and stator voltage.*

Thirdly, even with these assumptions of the equation remain extremely nonlinear and complex. In compiling the equations of elements of the vector control unit, additional simplifications are assumed, for example—the constancy of the rotor magnetic flux and the equality of the frequency of the stator voltage and the speed of rotation.

"The coordinate junction block (CJB) can be constructed on the basis of the equations of the control model controlled by voltage ([1], p. 2.22). They can be put $\omega_1 = \omega$ and $\frac{d\Psi_2}{dt} = 0$."

Thus, the generally accepted alternating current equations for vector control are largely simplified equations, which, in principle, incorrectly describe transient processes associated with changes in the frequency of stator voltage. Such changes lead to a change in the substitution schemes themselves, because their parameters: $x_1 = \omega_1 L_1; x_2 = \omega_1 L_2; x_m = \omega_1 L_m$ directly depend on the frequency.

The "qualitative processes" of the drive reaction on stepped load jumps can be obtained by combining the calculations of the substitution schemes with the calculations of the mechanical characteristics. If the frequency of the stator voltage does not change, then such descriptions are sufficiently accurate, since only the element $\frac{r_2}{S}$ is changed in such modes in the substitution schemes, and quite quickly. In this case, the initial and final states correspond to two substitution schemes with different $\frac{r_2}{S}$ and, respectively, by different current vectors. But with more complex frequency control algorithms, under which significant changes in the stator voltage occur—amplitude and frequency, new vector diagrams and substitution schemes are "formed" at each moment in which the frequency ω_1 changes. In these cases, high-quality analysis is significantly complicated, since all elements of the substitution scheme change, and the transition from one state to another vector equations is not described in principle. A description of such processes and their correction require a different approach or at least understanding the causes of the problems of existing approaches. This article provides several experiments, the results of which make it possible to understand the processes of stepped load jumps in asynchronous drives with different frequency control methods.

All equations for alternating current machines are designed in the 20s of the last centuries, when adjusting the frequency of the stator voltage was the problem of a distant future. In the past 30 years, this future has come, but simplified remains and make final control errors too significant, and theoretical provisions and even modeling are most often inconclusive. The most significant results of studies of asynchronous electric drives in this situation are experiments, and as close as possible to real industrial conditions.

At the same time, the experiments also require special justifications, since the processes in the asynchronous electric motor rotor are not available for measurement.

This state of affairs restrains the introduction of asynchronous electric drives in new areas for them—in aggregates requiring speed and accuracy. In addition, it is difficult to optimize electric drives in power engineering and transport where they are widely used. At the same time, their economy, acceptable price and high reliability remain a significant cause of research to improve their controllability.

Consider the main generally accepted methods for control of asynchronous electric drives.

2. Formulation of the problem

2.1 Physics of traditional control methods

Scalar control, which is customary to be simple and reliable—at a given speed, the scalar is selected—amplitude and frequency of change of voltage supplied to the engine stator. Mechanical characteristics are determined by the properties of the electric motor and the dependence $U(f_1)$. The structural scheme in **Figure 2**.

In **Figure 3a** shows the diagrams of the active values of the stator current and speed during acceleration and when the load is signed, obtained during stand-based studies, described in detail several articles [5–7].

The transition trajectory is determined by processes in the motor. Qualitatively these processes are as follows: When exposed to the load torque, the speed of rotation decreases, currents in the rotor and the stator increase, the slip in the motor grows and increases the torque developed by the motor to the state to which the torque corresponds equally to the load. If the parameters of the motor and the substitution schemes are "correct," the process occurs without oscillations and is fast enough, as in the examples of experiments in **Figure 3a**.

In **Figure 3b**, the mechanical characteristics of the motor and the trajectory of the transition from point A (with low load) to point B (with a large load torque).

In **Figure 3c**—the mechanical characteristics of the transition to the mechanical characteristic, corrected IR-compensation during load. Changes in working points B_1 and B are minor.

2.1.1 Vector control sets the task of efficient drive control

To do this, in the models embedded in the control unit of the FC on the measured values of the stator current and stator voltage, the required parameters of the stator voltage vector are calculated, which may vary at any time. The initial engine equations—significantly nonlinear undergo many simplifications, the main of which are the constancy of the rotor flow, the equality of the frequency of the stator speed voltage and the absence of other harmonics [2].

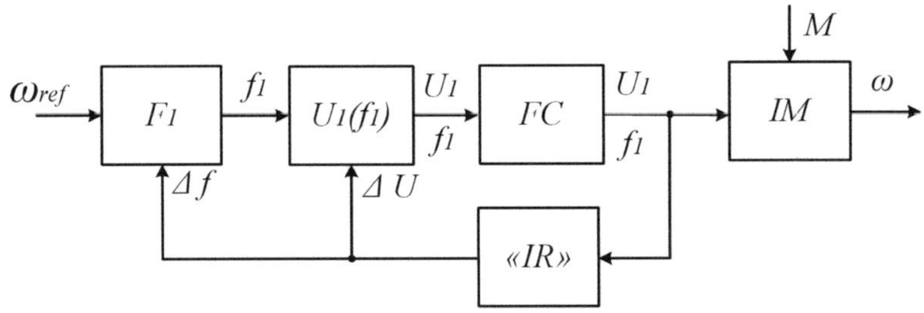

Figure 2.
Structural scheme of asynchronous electric drive with scalar control.

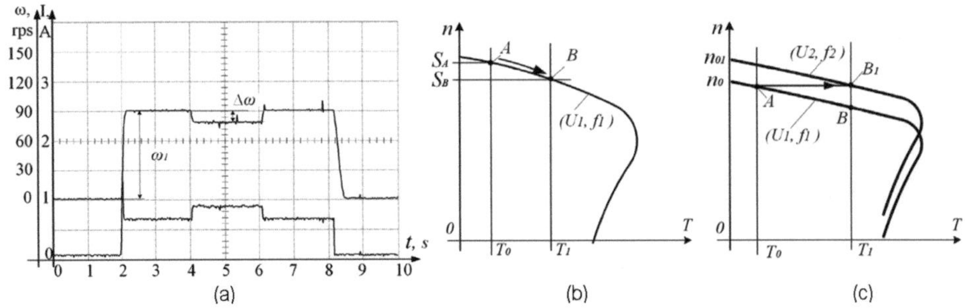

Figure 3.
Process diagrams (a) and mechanical characteristics (b and c) of asynchronous drive with scalar control.

$$\omega_1 = \omega; \frac{d\Psi_2}{dt} = 0 \tag{2}$$

In this case, the algorithm for the control of vector sensorless control picks up the vector of stator voltage, but not the transition path from one vector to another (**Figure 4**).

(Transition trajectories are given special attention in a separate method of vector control—"Direct Torque Control." This method is used mainly by ABB. In the articles dedicated to the method, the algorithms are described mainly at the level of logical provisions. As part of this work, this method of control will be a dedicated comment. Special studies have not been conducted.)

The purpose of the vector control algorithms is to linearize the drive and bring its characteristics to the characteristics of the DC drive characteristics, but the assumptions and errors adopted at these conclusions, as well as the inconsistency of the reality model makes the control error. It should be noted that for linearization, the vector control uses serial corrective devices that do poorly perform these functions. Especially in variations in the characteristics of the linearizable object. "Interferes" into operation and correction and impulse nature of the power of the frequency converter, through which the "correction" of the nonlinearities of the asynchronous electric motor occurs. These features of vector control are noted by almost all researchers [1–4]. In general, in the opening vector control of the change of stator voltage, the load jump is not too large and the processes of the load jump are not much different from the scalar control—**Figures 3c** and **5b**.

Experiments have shown that when overclocking the task signals with a small and unchanged load, assumptions (1) can be considered permissible and the engine equations quite correctly take into account changes in the frequency of the stator voltage and form the correct transition path from the vector state with one frequency to another, but in parrying mode adaptation loads does not occur, the differences in the speed of rotation and frequency of the stator voltage is a significant amount (absolute slip). And the transition trajectory to another state is not corrected. Most often, vector sensorless control poorly adjusts the voltage parameters on the stator and the engine is steaming the load in the same way as with a scalar control mode (**Figure 5a**). Mechanical characteristics are shown in **Figure 5b**. Vector control eliminates unstable "branches" of mechanical characteristics, at the same time, work areas differ little with a scalar. In **Figure 5b** shows the mechanical characteristics of the drives with the reaction to the load diagram and the transition trajectory.

2.1.2 Vector control with a speed loop

By analogy with direct current drives, the additional linearization circuit should carry out the speed control circuit with the PID regulator.

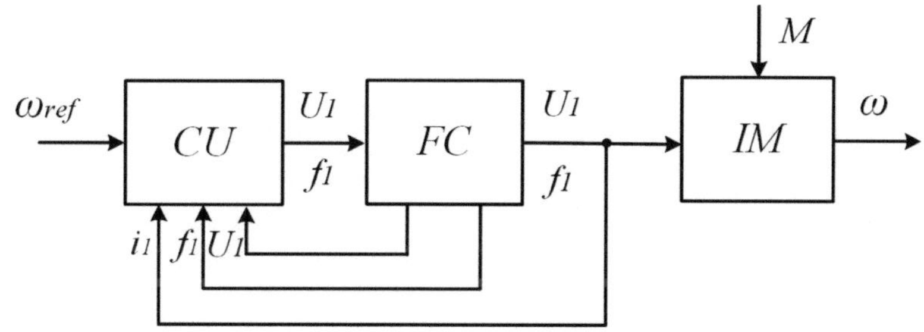

Figure 4.
Structural scheme of asynchronous electric drive with vector sensorless control.

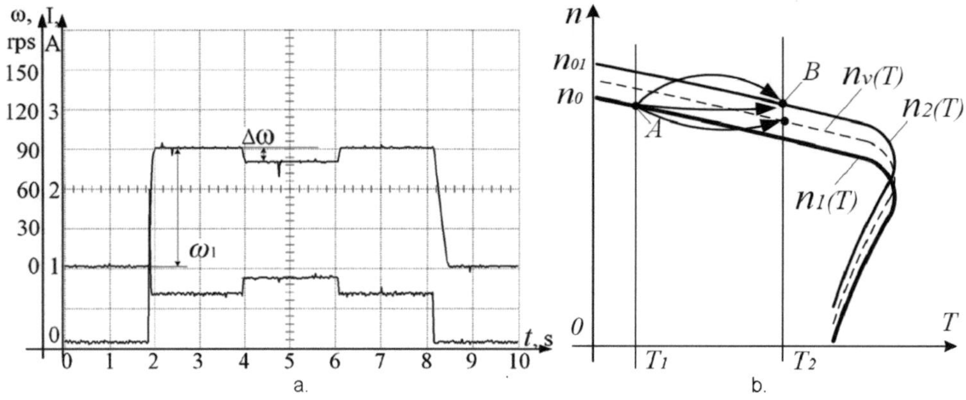

Figure 5.
Process diagrams (a) and mechanical characteristics (b) of asynchronous drive with vector control.

Speed sensors are quite rarely installed on general industrial mechanisms as shown experiments on the stand of a special effect on their application in drives with vector control is not too significant.

When the rotation speed circuit, the control signals for the FC are formed in the PID controller, the inputs of which are sent to the speed of rotation speed and the feedback signal. At the same time, all the problems of dumbfounded vector control are only aggravated (**Figure 6**).

One of the main assumptions of vector sensorless control is the equality of the frequency of the stator voltage and the speed of rotation in the output of the equation coordinate junction block (CJB) of the vector control ([1], p. 61):

$$\omega_1 = \omega \tag{3}$$

When controlling the speed of rotation on the side of the task with the PID controller, this is quite acceptable, but when the loading torque is parried, the permissibility of speed equality and frequency is fundamentally incorrect.

At the time of the jump of load, a dynamic failure of the motor speed occurs, the output of the speed controller generates a signal to an increase in the frequency of the rotor voltage following the conditions (assumptions) in the drive, this leads to a mitigation of the mechanical characteristic (**Figure 7c**), an increase in the dynamic rate of speed and tighten the speedy recovery. This follows from the process diagrams and the transition paths on the graphs of mechanical characteristics (**Figure 5**).

"Double" linearization with very significant assumptions leads to the fact that the acceleration of the drive is similar to the acceleration of the DC drive, and the

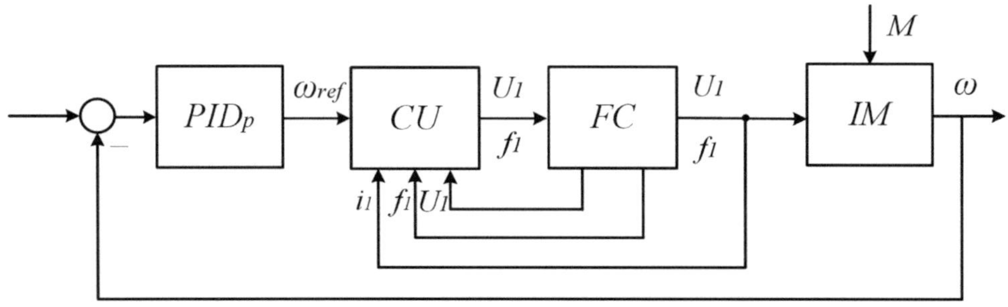

Figure 6.
Structural scheme of asynchronous electric drive with vector sensorless control and speed loop.

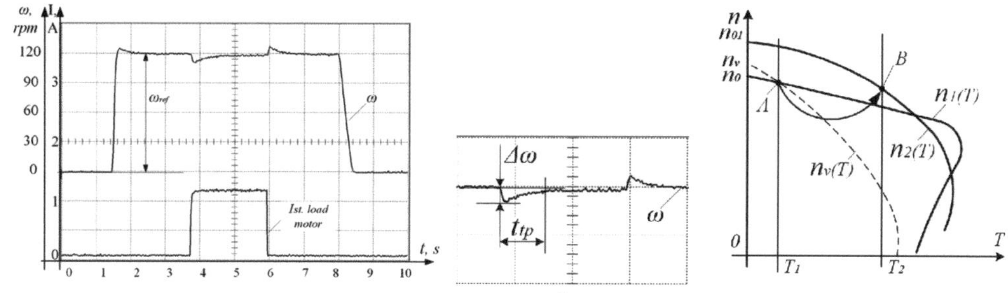

Figure 7.
Process charts (a and b) and mechanical characteristics (c) asynchronous drive with vector control and PID-regulator on speed loop.

processes of parrying the torque load in such a drive have zero static error in speed. However, the efficiency of such a drive cannot be considered significant. The time of the transition process and the dynamic failure of the speed are such that it is impossible to use this option of parrying the load for almost any industrial mechanism.

Comparison of the process of parrying of torques shows that the initial and final states can be the same in the drive, but the transitions between them have an infinite set of trajectories, which are determined by the stator voltage, and the control method form the transition path and the final vector.

2.1.3 Mathematics describing modes of operation

A significant role in the formation of algorithms is played by the mathematics of the description of processes in alternating current electric machines. In describing the operation of the SCIM [2], vector equations and dependencies with a large number of assumptions and simplifications are used.

Vector equations describing all asynchronous and synchronous motors, do not take into account the variable nature of the frequency of voltages and currents. It should be recognized if you assume that the frequency of the stator voltage is a complex function of time, the transition from Eqs. (2.19) to (2.21) ([1], p. 56) will be impossible and, the motor equations are complicated so much to analyze them and choose them effective correction will be impossible.

In the works [5–11], a nonlinear transfer function was proposed linking the mechanical torque developed by the SCIM and the absolute slip—the difference between the frequency of the stator voltage and the speed of rotation of the engine. The formula of this function includes, as variables, the frequency of the stator voltage and the relative slip. The formula can be called a nonlinear transfer function or a dynamic Kloss formula. In the articles [5–8], the conclusion of the proposed nonlinear transfer function is given in sufficient detail, the result is as follows:

$$W(p) = \frac{2M_k\left(T_2'p + 1\right)S_k}{\omega_1\left[\left(1 + T_2'p\right)^2 S_k^2 + \beta^2\right]} \tag{4}$$

where, ω_1 is the frequency of stator voltage, β is a relative slip, depending on the load of the drive.

This gear ratio corresponds to the structural diagram of the SCIM, shown in **Figure 8**.

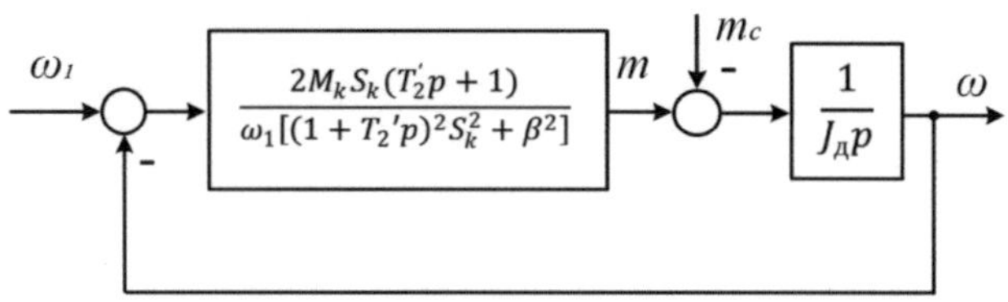

Figure 8.
Block diagram of the ADCZ with a nonlinear transfer function of the link forming the torque.

The works [9–13] show how it is possible to linearize the specified transfer function, that is, to exclude or significantly weaken the dependence of the transfer function and the dynamics of the asynchronous drive on the frequency of the stator voltage and sliding by positive feedback on the developed torque. The block diagram will take the form (**Figure 9**):

The transfer function of the corrective link, which is necessary for positive feedback to maintain the stability of the drive, is as follows:

$$W_{\text{DPF}} = \frac{\omega_1 \beta^2}{2M_k S_k \left(T_2' p + 1\right)}; \tag{5}$$

The equivalent transfer function of the drive with this connection will take the form:

$$W_{\text{eqv}} = \frac{2M_k S_k \left(T_2' p + 1\right)}{\omega_1 \left[(1 + T_2' p)^2 S_k^2\right]} = \frac{2M_k}{\omega_1 S_k (1 + T_2' p)} \tag{6}$$

From the point of view of mathematics, the transfer function with parameters depending on the functions of frequency and slip, this is the same inaccurate mathematical expression, as well as the vector equation, originally derived at unchanged frequencies of harmonic variables—the currents of the engine and EMF rotor and the stator used to analyze the dynamics of these same variables. However, there is one significant difference. If the vector equation is valid exclusively for the constant frequencies of signals associated with the equation, and in principle cannot describe the change in these frequencies, the transfer function retains its ability to describe the dynamics of processes in some area of changes in these functions and even has sufficient accuracy of this description.

So, when the drive is working out a torque disturbance at a constant rotational speed (and a constant frequency of the stator voltage) with a slight (for the transfer function) change in the relative slip β, the transfer function describes the processes quite accurately, and, more importantly, allows you to accurately select corrective connections that linearize the transfer function and make the parrying of the

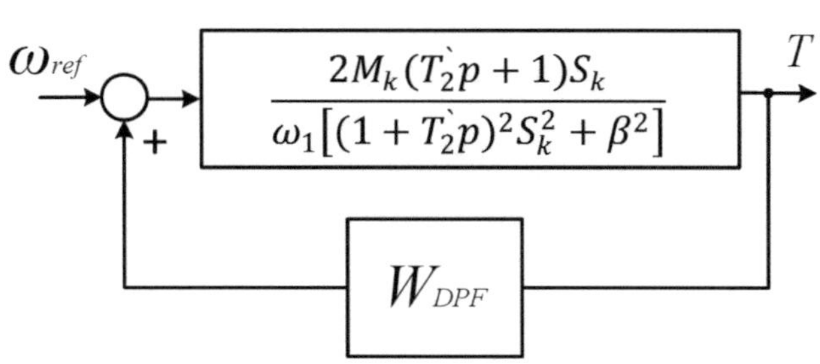

Figure 9.
Block diagram of SCIM with dynamic positive feedback (DPF).

disturbance in the drive much more efficient. So, the proposed positive dynamic feedback by the torque of the engine or its analogue:

$$W_{DPF} = \frac{K}{(T'_2 p + 1)} \tag{7}$$

The transfer function forms and optimally the transition trajectory and significantly reduces the time of transient processes. Dynamic "dips" speeds are also reduced. Experiments exploring the reaction of the drive to jump of load at various methods of controlling the SCIM fully confirmed this (**Figure 10b**).

The processes in **Figure 10b** show that DPF allows correcting the static error of the drive and significantly speeding up transients. At the same time, the nonlinear transfer function can describe transients and justify continuous devices for their correction, which is the positive feedback on the active component of the stator current. The stator voltage "selected" by this feedback (voltage amplitude and frequency) provides parrying of step loads with minimal transient processes (**Figure 10b**).

This made it possible to formulate a hypothesis that the identification of SCIM with FC a non-linear transfer function is more accurate than vector equations, which is confirmed by the choice of a more effective correction. In asynchronous electric drives using widely used frequency converters, it is quite problematic to introduce positive torque feedback. As experiments have shown [14, 15], it can be replaced by a connection according to the active component of the stator current, which is measured by almost all known frequency converters used in industry.

As mentioned above, the crucial importance in assessing the correctness and effectiveness should be given to experimental studies.

The stand where the research was carried out initially consisted of two identical asynchronous electric drives, each of which contains an asynchronous

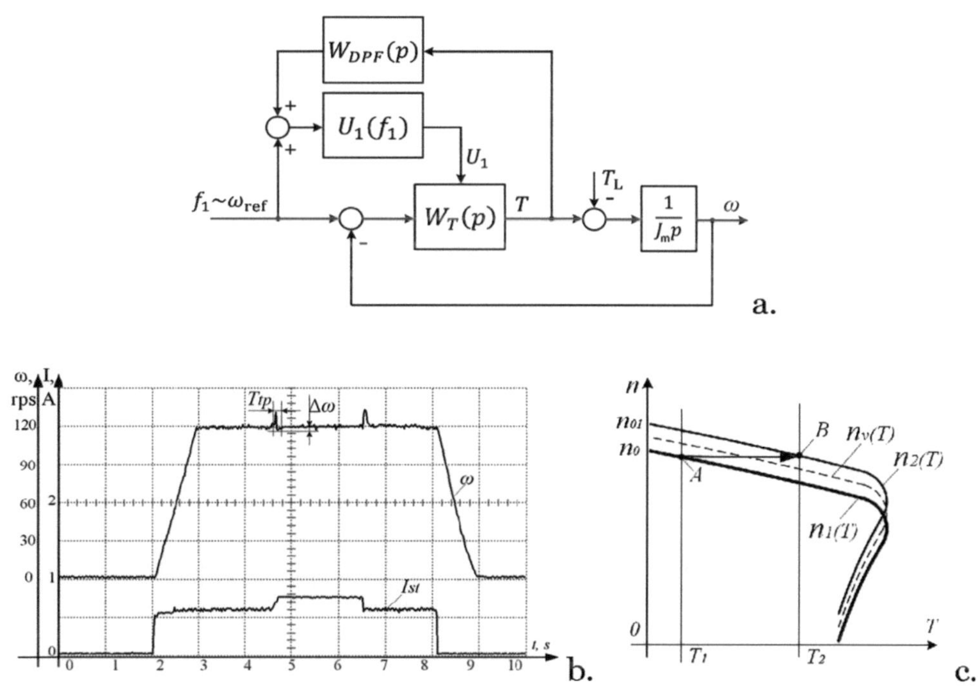

Figure 10.
Structural diagram of the actuator with dynamic bond on time (a), transient processes (b), and mechanical characteristics (c).

short-circuited electric motor and a frequency and voltage converter. The drives operate on one common shaft, the stand contains current sensors and a rotation speed sensor of the common shaft of the motors and a generator of periodic control signals. Quite a lot of different experiments were carried out, described in detail in the articles [16–20].

This study provides the results of experiments during steady-state after the jump of load. These modes of operation are selected, the most exactly corresponding to the vector equations of asynchronous electric drives and well-specific qualitative analysis using well-known methods, mechanical characteristics of the drive (**Figure 11**).

The technology of experiments is extremely simplified. A certain control mode is set in the working drive—vector sensorless or with speed feedback, scalar, or with positive feedback—DPF or DPF2 (DPF2 is a dynamic positive feedback, similar to DPF, but with an increased transmission coefficient (Kt = 3)). Scalar control is installed in the load drive. Directly by the signal supplied to the input of the frequency converter *UZ2* of the working motor *M2*, the drive is output to a certain rotation speed (and the corresponding frequency of the stator voltage). After a certain time interval, a task is sent to the frequency converter *UZ1* of the load motor *M1*. The operating mode of the load drive is determined by the task for the speed of rotation, as an equivalent of the mechanical characteristic (**Figure 12**). The drives work counter. The resulting modes are well explained by the mechanical characteristics—**Figure 12**. The working points are determined by the intersection of the mechanical characteristics of the working and load motors. At the same time, **Figure 12a**—U/f = const., **Figure 12b**—$U_2/f_2 > U_1/f_1$.

The jump of load smoothly enough, the rate of load increase is commensurate with the processes of torque formation in the working drive.

The parameters of the modes—the stator currents and the rotation speed (or the sliding value) are determined by the vectors of the stator voltage, which the corresponding control algorithm will “choose.” The diagrams of the speed and current of the stator are similar to those shown in **Figures 4–6**.

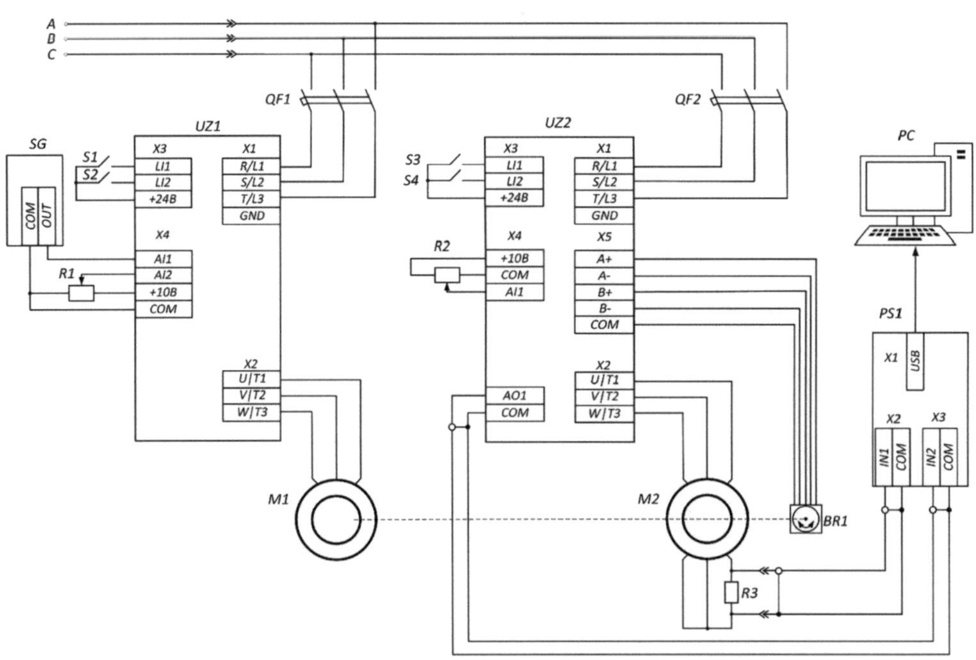

Figure 11.
Stand scheme.

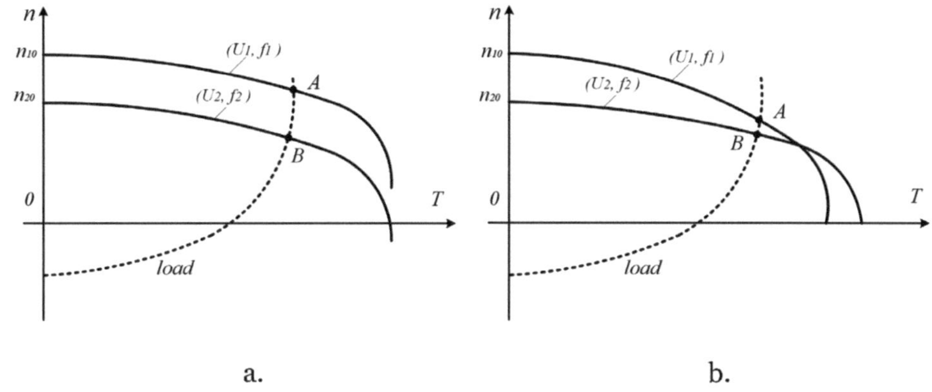

Figure 12.
Mechanical characteristics of the counter activation of the working drive of the stand with different parameters of the stator voltage (U, f) *and the load drive. (a)* U/f = *const. and (b)* $U_1/f_1 > U_2/f_2$.

As follows from the figures, it is quite difficult to evaluate the efficiency of torque generation in the drive-by one or another control method using these diagrams. Since all these algorithms control the frequency of the stator voltage to a greater or lesser extent, even by the speed signal, it is difficult to estimate the sliding in the motor during experiments.

A methodology was proposed for evaluating the effectiveness of the AED control method by sliding, necessary for the formation of torque in the motor.

Sliding can be determined most accurately in a real drive by the frequency of the rotary current. To work with this technique, the working motor in the stand was replaced with an electric motor with a phase rotor, in which rotor current sensors were installed. **Figures 13** and **14** show diagrams of rotor currents in the working motor of the stand in a circuit with dynamic feedback, and in a circuit with vector control and a PID speed controller, respectively. The results were very telling. The frequency of the main harmonic of the rotor current in the drive with DPF (3.5 Hz) is significantly lower than in the drive with a PID speed controller (8.125 Hz).

In the analysis of the experiments, the main attention was paid to the frequency of rotor current, which is in the circuit with the DPF was significantly lower. But also turned out to be smaller and the amplitude values of the rotor current. To carry out a more detailed assessment of the effectiveness of choice of the stator voltage vector, new experiments were conducted.

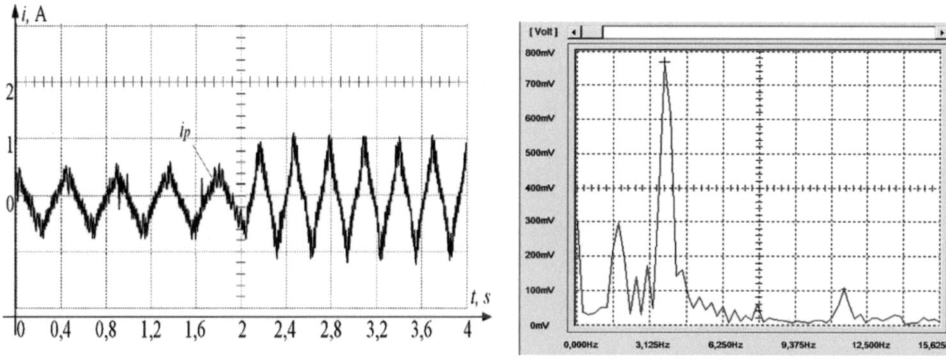

Figure 13.
Rotor currents and the spectrum of rotor current in drive with scalar control and DPF.

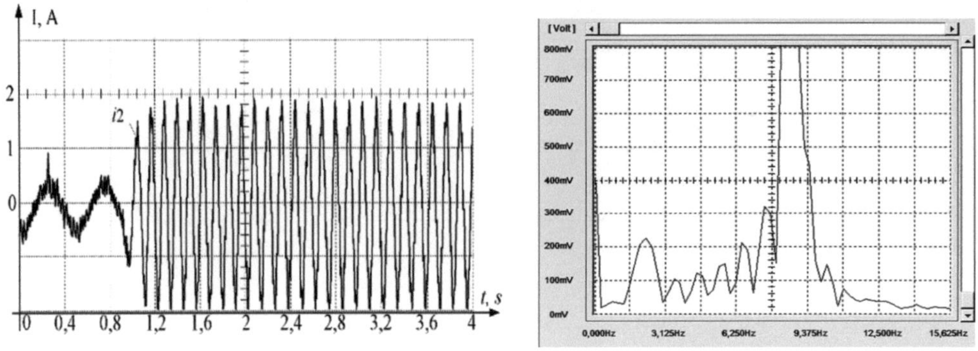

Figure 14.
Rotor currents and the spectrum of rotor current in drive with vector control and speed loop.

3. New experiments with static modes

3.1 Methodology and course of experiments

The working drive, in which five control methods are implemented sequentially, is output at rotational speeds corresponding to the set frequencies of the stator voltage—20 and 30 Hz—**Table 1**.

The load drive generates a braking torque corresponding to the "counter" mechanical characteristic with a given rotational speed of 10 and 15 Hz.

Signals of currents and frequencies are recorded in the rotor and stator of the working drive without load and when switched on with the load drive in steady-state modes. Data in **Tables 1–5**.

The values of the voltage U_1, the amplitude of the stator current I_1, the rotation speed recalculated in FC relative to a given frequency are recorded according to the

	F_1, Hz	U_1, V	Fs, Hz	I_1, A	F_2, Hz	I_2, A	F_1	U/f_1
SC	20	91	18.6	0.5	1.5	1.48	20.1	4.5
VC	20	83	18.1	0.4	1.75	1.88	19.85	4.3
SVC	20	90	20	0.39	1.72	1.76	21.76	4.3
DPF	20	92	18.6	0.42	1.42	1.3	20	4.5
DPF2	20	103	19.7	0.36	1.3	1.3	21	4.9

Table 1.
Parameters of the working drive at low load, the load drive is switched off.

	F_1, Hz	U_1, V	Fs, Hz	I_1, A	F_2, Hz	I_2, A	F_1	U/f_1
SC	20	91	6.1	0.7	13.8	12.8	19.9	4.5
VC	20	84	0	0.8	20	15.8	20	4.2
SVC	20	111	11.8	0.7	14.9	14.6	26.7	4.2
DPF	20	103	10.6	0.7	12.2	11.3	22.8	4.6
DPF2	20	110	12.9	0.7	9.6	10	22.5	4.9

Table 2.
Parameters of the working drive at a load of 70% of Mn *(setting the speed of the load drive is equivalent to setting "15 Hz").*

	F_1, Hz	U_1, V	Fs, Hz	I_1, A	F_2, Hz	I_2, A	F_1	U/f_1
SC	30	135	28.8	0.5	1.5	2.1	30.3	4.5
VC	30	135	29	0.5	1.1	2.5	30.1	4.5
DPF	30	134	29	0.5	0.9	2.6	31.6	4.4

Table 3.
Parameters of the working drive at low load, the load drive is switched off.

	F_1, Hz	U_1, V	Fs, Hz	I_1, A	F_2, Hz	I_2, A	F_1	U/f_1
SC	30	138	24.8	0.6	7.5	6.2	32.3	4.2
VC	30	138	24.7	0.6	7.0	7	31.7	4.3
DPF	30	152	28.3	0.6	6.6	6.8	35.1	4.3

Table 4.
Parameters of the working drive at a load of 50% of Mn *(setting the speed of the load drive is equivalent to setting "10 Hz").*

	F_1, Hz	U_1, V	Fs, Hz	I_1, A	F_2, Hz	I_2, A	F_1	U/f_1
SC	30	139	21.6	0.7	10	9.2	31.6	4.4
VC	30	139	21.8	0.7	9	9.2	30.8	4.45
DPF	30	165	29	0.7	9	9.6	38	4.3

Table 5.
Parameters of the working drive at a load of 70% of Mn *(setting the speed of the load drive is equivalent to setting "15 Hz").*

readings of the FC monitor, the rotor current is recorded by the Instek GDS-2062 oscilloscope.

The actual values of the frequency of the stator voltage *Fs* and the ratio U/f_1 in each experiment are calculated from the values of the measured rotational speed and frequency of the rotor current. These relations determine, as indicated in [6, 10], the main magnetic flux in the engine, the value of the critical torque and, according to the Kloss formula, the slip required to create torque in this static mode. The frequency of the rotor currents determines the actual slip in the working drive.

Comment 1: The stator current changes little by increasing the load and the torque of the "working" electric motor—0.5 A—without load, and 0.6–0.7 at during loads, its changes cannot be analyzed. Rotary current varies significantly—from 1.3 to 2.6 A without load and from 6 to 16 A with loads.

It is a rotary current, being active, creates a torque. DPF communication, managing simultaneously and frequency, and voltage—*U/f*, retains its value and the magnitude of the main magnetic flux in the engine, the value of an absolute slip and the amplitude of the rotor currents also change little when the control algorithms change. But the speed of rotation under load is better adjusted than in drives with "open" algorithms. The DPF connection simply "shifts" the mechanical characteristics parallel to natural—**Figure 12a**.

Communication DPF2 increases the *U/f* and at the same time—the main magnetic flux in the motor. The drive currents, sliding and the reactive power flows are reduced, as compared with the electric drive with DPF—**Figure 12b** and **Tables 6** and 7. Since these control algorithms are formed, as seen from the tables, different stator voltage vectors to establish a clear connection between the control algorithm,

F_1, Hz	U_1, V	Fs, Hz	I_1, A	F_2, Hz	I_2, A	F_1	U/f_1
30	137	28.4	0.432	1.695	1.8	30.1	4.6
30	150	28.6	0.472	1.351	1.88	29.95	5
30	180	29.3	0.576	0.9	1.76	30.2	6
30	200	29.6	0.680	0.75	1.72	30.35	6.8
30							

Table 6.
Parameters of the working drive at light load with only scalar control, the load drive is off.

F_1, Hz	U_1, V	Fs, Hz	I_1, A	F_2, Hz	I_2, A	F_1	U/f_1
30	137	24	0.480	5.8	5.8	29.8	4.6
30	150	25.2	0.496	5	5.44	30.6	5
30	180	26.7	0.600	3.3	5.12	30	6
30	200	27.4	0.680	2.63	4.48	30	6.8

Table 7.
Parameters of the working drive with only scalar control, at a load of 50% of Mn (setting the speed of the load drive is equivalent to setting "10 Hz").

this vector and the drive mode at loads, that is, values of currents and their frequencies have been carried out in additional experiments.

Additional experiments were carried out in the following order.

In the scalar control mode, a different amplitude of the stator voltage from 130 to 200 V was set for a certain rotation frequency. All process parameters in the stator and rotor were recorded at low load (**Table 6**). The parameters of the processes in the drives when the load drive is turned on, rotating in the opposite direction, are given in **Table 7**.

Comment 2: The analysis showed that when the amplitude and frequency of the voltage across the motor stator are set similar to the parameters of the "selected" DPF and DPF2 modes, the parameters of the processes in the rotor and stator turn out to be the same with scalar control.

At the maximum value of U/f = 6.8, the minimum values of the rotor current (1.72 A) and slip (4.48 Hz) under load, and 0.75 A and 2.63 Hz, respectively, at low load are fixed.

This confirms the assumption that it is the parameters of the stator voltage (U, f), which determine the main magnetic flux in the motor, that determine the operating mode of the AED—sliding and rotor currents. The larger the main flow, the smaller the slip and the rotor current.

The control algorithms (SC, SVC, DPF, and DPF2) only "select" the values of U and f according to the task for the rotation speed, the load torque and in accordance with the control algorithm.

The DPF control algorithm relies on a continuous nonlinear transfer function—a more accurate interpretation of the AED, since it does not have the assumptions that are made in vector control (only the main harmonics in the currents and EMF of the motor, $\omega_1 = \omega$, $\frac{d\psi_2}{dt} = 0$), "selects" control more efficiently.

This correction method is certainly more promising both for static and quasi-static modes, and for operation under complex disturbances and in complex technological systems (special vehicles, wind turbines, drones, technological complexes, power engineering ...).

3.2 The discussion of the results

As experiments have shown, the static modes of operation of the AED, given U and f, can be accurately described by vector diagrams and mechanical characteristics. At the same time, it is necessary to clarify the amount of slip required to create torque.

The "selection" of the values of the amplitude and frequency of the stator voltage is carried out by the control algorithm—SC, SVC, and DPF. With an increase in the main flux in the motor, the stator current at low load increases, while the rotor current and slip at high load fall. That is, positive torque feedback provides low no-load stator currents and reduced rotor currents under load.

Traditional control algorithms (SC, SVC) do not choose U and f optimally, since the algorithms are based on vector equations and several assumptions that are very erroneous for working under load even in static modes ($\omega_1 = \omega$, for example). The control transition paths are selected incorrectly.

The use of continuous nonlinear transfer functions interpreting AED and corrections by local feedbacks on these functions form transition processes (including transition trajectories) in complex operating modes with variable load.

This determines their advantages and prospects for the use of continuous transfer functions and continuous local corrections and structures in AED complex technological complexes.

This is especially important because, in terms of price, reliability and overload capacity, AED has no alternatives.

3.3 About vector control

Theoretical "understanding" of vector control problems shows that sequential correction, which is vector control (**Figure 3**) with simplifications of the original equations, passing through nonlinear blocks with delays (the pulsed power part of the FC) is very inefficient and experiments confirm this, although statics is closest to the original vector equations. We should expect even greater problems in operating modes with significant dynamics of loads and rotational speeds. The sequential correction that vector control "tries" to implement does not work, because it is based on many erroneous simplifications and is ineffective as a sequential correction of the nonlinear structure of torque generation in asynchronous motors.

3.4 About direct torque control (DTC) technology

Over 30 years of direct torque control (DTC) technology, a lot of work has been devoted. An example of the scheme is shown in **Figure 15**.

Most often, there are no detailed descriptions, there is no technology, including the characteristics of the "observer" of flow coupling—its accuracy, dynamics, which fundamentally affect all processes. Another obvious, but ignored in the descriptions of technology, the problem is the presence of high harmonics in all. This "suggests" some kind of conditionality of the results.

The very appearance of this kind of vector control is probably a reaction to the poor-quality operation of the vector control at the entrance from the torque of the disturbance and represents the formation of a transition trajectory from one vector state to another, improved compared to the vector transition, in which this trajectory is not paid attention at all. The trajectory is carried out at the expense of "basic" vectors. The algorithm software is very complex and other companies (except ABB) do not use it. Confirms that discontinuous vector control does not give trajectories of transitions from one state to another.

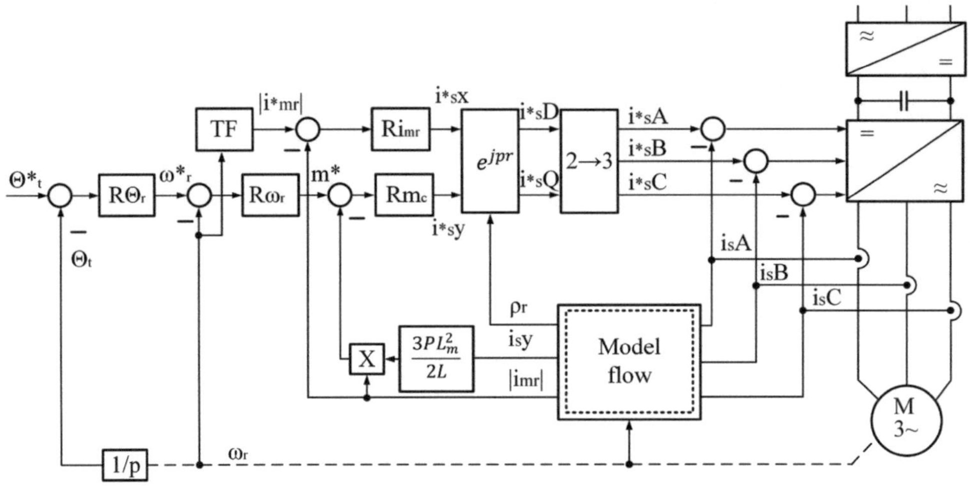

Figure 15.
Block diagram of direct torque control (DTC) in an asynchronous electric drive.

On the speed of the formation of the torque, which is often mentioned in the description of technology [21–24]. The transition time is 1–2 ms without specifying the drive power and specific charts are difficult to consider a convincing argument. In the article [21] the increasing time of 50 ms is also fast, but already real. In our experiments (**Figure 10a**) The process of recovery of speed is 80–100 ms. That for the torque commensurate with DTC. It is implemented in any FC and has prospects for improvement.

It should be noted that in the articles to directly control the torque [21–24] there are no at least some mathematics of the description of the dynamics—differential equations, etc.

On local connections based on the representation of an asynchronous electric motor by a nonlinear transfer function.

The transfer function of the link forms the torque in an asynchronous electric motor—the initial and adjusted ones are determined by the transition conditions, that is, the initial and final conditions for changing the load and frequency of the stator voltage.

The continuity of the transfer function contributes to optimizing the formation of the transition trajectory from one state to another. With the original vector equations, this transition is not determined and an additional algorithm is required—not only the basic vectors, but also the trajectories of transitions to them.

Since in all experiments, that is, at different operating speeds and loads with a deep positive connection in the stator current (more precisely, in the developed torque of the sliding value), the amplitudes of the currents are smaller than with traditional control methods (scalar and vector), it can be reasonably argued that the amplification of the main magnetic flux in asynchronous motors reduces the angle of shift between the reduced vectors of the rotor and stator currents of the motor.

4. Conclusion

The static modes of operation of the AEP in the control methods under consideration are described fairly accurately by nonlinear transfer functions (NTF), but transitions between static states are described incorrectly in these algorithms and when loading the work of these algorithms—in closed vector circuits—is incorrect and leads to significantly inefficient modes.

The local connection formed by the NTF is a positive dynamic feedback, "selects" the parameters of the stator voltage (U, f) the best for parrying static disturbing torques.

Discontinuous vector AED equations, based on which correction algorithms are formed by traditional methods, do not allow to obtain optimal modes of parrying load surges with their help.

The advantages in the quality of operation of control systems using the proposed local connections are even more obvious with large variable external disturbances and when working in complex ACS.

There is a proposal to introduce into the drive a positive feedback by the amplitude of the rotor current, as an analogue of the feedback by torque.

The operation of AED in complex systems and under complex disturbing influences is not described in principle (with sufficient accuracy) by the method of vector equations, hence the endless modes of selection and calculation of engine and drive parameters, automatic identification, and, finally, DTC technology.

Feedback by the torque or by the value of the active current of the stator under load reduces the currents of the stator, rotor and slip, that is, makes the processes more active, which is extremely "useful" for drives with large and varying loads.

Author details

Vladimir L. Kodkin*, Alexandr S. Anikin, Alexandr A. Baldenkov and Natalia A. Loginova
South Ural State University, Chelyabinsk, Russian Federation

*Address all correspondence to: kodkina2@mail.ru

References

[1] Usoltsev AA. Vector Control of Asynchronous Motors: Tutorial. Spb.: ITMO; 2002. p. 120. Available from: http://servomotors.ru/documentation/frequency_control_of_asynchronous_motors/chastupr.pdf

[2] Park R, Robertson B. The reactances of synchronous machines. Transactions of the American Institute of Electrical Engineers. 1928;**47**(2):514-535

[3] Vas P. Vector Control of AC Machines. New York, NY, USA: Oxford University Press; 1990

[4] Mishchenko VA. The vector control method of electromechanical converters. Electrical Engineering. 2004;7:47-51

[5] Kodkin VL, Anikin AS, Baldenkov AA. Identification of AC drivers from the families of frequency characteristics. Russian Electrical Engineering. 2020; **91**(12):756-760. DOI: 10.3103/S1068371220120081

[6] Kodkin V, Anikin A. On the physical nature of frequency control problems of induction motor drives. Energies. 2021; **14**(14):4246. DOI: 10.3390/en14144246

[7] Kodkin VL, Anikin AS, Baldenkov AA. The dynamics identification of asynchronous electric drives via frequency response. International Journal of Power Electronics and Drive Systems. 2019;**10**(1):66-73. DOI: 10.11591/ijpeds.v10n1.pp66-73

[8] Kodkin VL, Anikin AS. Experimental study of the VFD's speed stabilization efficiency under torque disturbances. International Journal of Power Electronics and Drive Systems. 2021;**12**(1):80-87. DOI: 10.11591/ijpeds.v12.i1.pp80-87

[9] Kodkin V, Baldenkov A, Anikin A. A method for assessing the stability of digital automatic control systems (ACS) with discrete elements. Hypothesis and simulation results. Energies. 2021; **14**(20):6561. DOI: 10.3390/en14206561

[10] Kodkin VL, Anikin AS, Baldenkov AA. Stabilization of the stator and rotor flux linkage of the induction motor in the asynchronous electric drives with frequency regulation. International Journal of Power Electronics and Drive Systems. 2020;**11**(1):213-219. DOI: 10.11591/ijpeds.v11.i1.pp213-219

[11] Codkin VL, Anikin AS, Baldenkov AA. Assessing the efficiency of control systems of asynchronous electric drives using spectral analysis of rotor currents. Russian Electrical Engineering. 2021; **92**(1):32-37. DOI: 10.3103/S106837 1221010065

[12] Kodkin VL, Anikin AS, Shmarin YA. Dynamic load disturbance correction for alternative current electric drives. In: 2016 2nd International Conference on Industrial Engineering, Applications and Manufacturing (ICIEAM): Proceedings: South Ural State University (National Research University), Chelyabinsk, Russia. 2016. DOI:10.1109/SIBCON.2015.7146978

[13] Kodkin VL. Methods of optimizing the speed and accuracy of optical complex guidance systems based on equivalence of automatic control system domain of attraction and unconditional stability of their equivalent circuits. In: Proceedings of SPIE—The International Society for Optical Engineering. 2016

[14] Kodkin VL, Anikin AS. Frequency control of asynchronous electric drives in transport. In: 2015 International Siberian Conference on Control and Communications (SIBCON 2015)—Proceedings. 2015. DOI: 10.1109/SIBCON.2015.7146978

[15] Kodkin VL, Anikin AS, Shmarin YA. Effective frequency control for

induction electric drives under overloading. Russian Electrical Engineering. 2014;**85**(10):641-644. DOI: 10.3103/S1068371214100101

[16] Kodkin VL, Anikin AS, Baldenkov AA. Experimental research of asynchronous electric drive with positive dynamic feedback on stator current. In: 2017 International Conference on Industrial Engineering, Applications and Manufacturing (ICIEAM)—Proceedings. 2017. DOI:10.1109/ICIEAM.2017.8076179

[17] Kodkin VL, Anikin AS, Baldenkov AA. Spectral analysis of rotor currents in frequency-controlled electric drives. In: 2nd International Conference on Automation, Mechanical and Electrical Engineering (AMEE 2017)—Proceedings. 2017. DOI: 10.2991/amee-17.2017.26

[18] Kodkin VL, Anikin AS, Baldenkov AA. Families of frequency characteristics, as a basis for the identification of asynchronous electric drives. In: 2018 International Russian Automation Conference (RusAutoCon). 2018

[19] Kodkin VL, Anikin AS, Baldenkov AA. Analysis of stability of electric drives as non-linear systems according to Popov criterion adjusted to amplitude and phase frequency characteristics of its elements. In: 2nd International Conference on Applied Mathematics, Simulation and Modelling (AMSM 2017)—Proceedings. 2017. pp. 7-14. DOI: 10.12783/dtetr/amsm2017/14810

[20] Kodkin VL, Anikin AS, Baldenkov AA. The analysis of the quality of the frequency control of induction motor carried out on the basis of the processes in the rotor circuit. Journal of Physics Conference Series. 2018;**944**(1):012052

[21] Karandeev DY, Engel EA. Direct torque control of an induction motor using adaptive neurocontroller in conditions of uncertainty. Internet Journal Science. 2015;7(5):1-9. DOI: 10.15862/91TVN515

[22] Wang F et al. Advanced control strategies of induction machine: Field oriented control, direct torque control and model predictive control. Energies. 2018;**11**(1):120. DOI: 10.3390/en110 10120

[23] Alsofyani IM, Idris NRN. Simple flux regulation for improving state estimation at very low and zero speed of a speed sensorless direct torque control of an induction motor. IEEE Transactions on Power Electronics. 2016;**31**(4):3027-3035. DOI: 10.1109/TPEL.2015.2447731

[24] Toufouti R. Direct torque control for induction motor using intelligent techniques. Journal of Theoretical and Applied Information Technology. 2007;**3**(3):35-44